Think, Do, and Communicate Environmental Science

Many students find it daunting to move from studying environmental science to designing and implementing their own research proposals. This book provides a practical introduction to help develop scientific thinking, aimed at undergraduate and new graduate students in the earth and environmental sciences. Students are guided through the steps of scientific thinking using published scientific literature and real environmental data. The book starts with advice on how to effectively read scientific papers, before outlining how to articulate testable questions and answer them using basic data analysis. The Mauna Loa CO_2 dataset is used to demonstrate how to read metadata, prepare data, generate effective graphs, and identify dominant cycles on various timescales. Practical, question-driven examples are explored to explain running averages, anomalies, correlations, and simple linear models. The final two chapters and the epilogue provide a framework for writing, summarizing, and evaluating persuasive research proposals, making this an essential guide for students embarking on their first research project.

TARA IVANOCHKO is an associate professor of teaching and Director of Environmental Science in the Department of Earth, Ocean and Atmospheric Sciences at the University of British Columbia. She is also Academic Director of the UBC Sustainability Initiative. Her research focuses on marine palaeoclimatology and the reconstruction of environmental systems on timescales from 10,000 to 100,000 years. A passionate educator, she has also developed a multi-year curriculum to actively engage undergraduate students in authentic scientific experiences, and has incorporated learning portfolios and community-based learning into the Environmental Science degree.

Think, Do, and Communicate Environmental Science

TARA IVANOCHKO
University of British Columbia

CAMBRIDGE
UNIVERSITY PRESS

University Printing House, Cambridge CB2 8BS, United Kingdom

One Liberty Plaza, 20th Floor, New York, NY 10006, USA

477 Williamstown Road, Port Melbourne, VIC 3207, Australia

314–321, 3rd Floor, Plot 3, Splendor Forum, Jasola District Centre,
New Delhi – 110025, India

79 Anson Road, #06–04/06, Singapore 079906

Cambridge University Press is part of the University of Cambridge.

It furthers the University's mission by disseminating knowledge in the pursuit of education, learning, and research at the highest international levels of excellence.

www.cambridge.org
Information on this title: www.cambridge.org/9781108423458
DOI: 10.1017/9781108526104

First published 2021

A catalogue record for this publication is available from the British Library.

Library of Congress Cataloging-in-Publication Data
Names: Ivanochko, Tara, 1972- author.
Title: Think, do, and communicate environmental science / Tara Ivanochko, University of British Columbia, Vancouver.
Description: Cambridge, UK ; New York, NY : Cambridge University Press, 2021. | Includes bibliographical references and index.
Identifiers: LCCN 2020040231 (print) | LCCN 2020040232 (ebook) | ISBN 9781108423458 (hardback) | ISBN 9781108437578 (paperback) | ISBN 9781108526104 (epub)
Subjects: LCSH: Environmental sciences–Research. | Environmental sciences–Research–Methodology. | Environmental sciences–Research–Data processing. | Environmental literature. | Technical writing.
Classification: LCC GE70 .I93 2021 (print) | LCC GE70 (ebook) | DDC 550.72–dc23
LC record available at https://lccn.loc.gov/2020040231
LC ebook record available at https://lccn.loc.gov/2020040232

ISBN 978-1-108-42345-8 Hardback
ISBN 978-1-108-43757-8 Paperback

Contents

Prologue

Large environmental changes are occurring all around us. As the human population has grown, the demand for food, energy, water, living space, and amenities has also grown, as has the production of industrial wastes, sewage, agricultural run-off, and carbon dioxide – the monster of all waste products. These changes are not always obvious to the naked eye, and the lack of a clear understanding is often a major barrier to taking appropriate action. This is where the environmental scientist is needed: to shed light on the what, where, when, why, and how of environmental change.

When I was barely one year old, my father Charles David Keeling started making measurements of atmospheric carbon dioxide at the Mauna Loa Observatory on the Big Island of Hawaii. The goals were to answer a couple of simple questions: Is atmospheric carbon dioxide increasing? If so, are humans responsible?

In those early days, he often was concerned that his instrument, which measured carbon dioxide around the clock, might break down or give erroneous readings. But within three years, he already had the answer to his first question: Yes, carbon dioxide levels were increasing. And within about six years, he could clearly demonstrate that the rate of increase was significantly slower than would be expected based on the amount of CO_2 emitted into the atmosphere each year by the burning of fossil fuel. This extra CO_2 was presumably being taken up by land plants or by the oceans. But this only raised new questions: How important were the land and ocean separately as sinks for CO_2? What processes controlled these sinks? How might they change in the future? And of course: would the continuing rise in CO_2 cause the planet to warm up?

He also found something totally unexpected. CO_2 was not only just rising year on year but it was also going up and down with the seasons. This reflected the annual cycle of plant growth dominated by photosynthesis in the summer and respiration in the winter: as plants grow, they consume carbon dioxide from the air around them, and as they decompose, they release carbon dioxide back to the air. He was effectively the first person to see the planet, as a whole, breathing.

Atmospheric carbon dioxide is now being measured at dozens of sites around the world. The Mauna Loa record is now more than 60 years long, and the records from the other sites also span many decades. We also have time series of data that track changes in the pH of the ocean, ozone concentrations in major cities, hurricane (drought and flood) frequency and intensity, the amount of human debris in space, and temperature and weather conditions from around the planet. These measurements further confirm that the increases in atmospheric carbon dioxide are caused by human activities: burning fossil fuels for energy, industry, transportation, and food production. They also tell us that land plants and the oceans are both important sinks for the CO_2, and yes, Earth's climate is warming. But again, there are new questions. How bad will things get if we continue on the same path? What needs to be done to get control of the climate problem? How can we demonstrate progress?

The Mauna Loa record continues to be relevant. For example, one of the best measures of progress is whether the curve itself starts to bend downwards. When do we expect to be able to see this happening? How will we know for sure?

When you first look at environmental data, and time series in particular, you can be overwhelmed. Environmental datasets can look messy, dense, and unpredictable. But hidden in the jumble of numbers are important stories. My father contributed more than just the numbers he was generating; he thought long and hard about what the data were telling him, how to graph the data, and how to make sense of the numbers.

Like other scientists of his day, my father's goal was purely to gain knowledge: to discover and report about the natural world. He certainly didn't consider himself to be an environmental scientist; this term didn't yet exist. But in his quest to understand the world around him in the context of human activities, he was actually an archetypal environmental scientist.

In the last few decades, environmental science has become a major field of interdisciplinary science studying the influences of humans on the environment and the influences of the environment on humans. Environmental datasets that extend over time, like the one my father started at Mauna Loa in the 1950s, now provide rich resources that expand our knowledge of Earth processes, and help us control our future.

Professor Ralph Keeling
Scripps Institution of Oceanography

Acknowledgements

This book was inspired by my students. I am always moved at the end of term when my students reflect on their accomplishments and express pride in their final research proposals. This book reflects what I have learned from students' interests, questions, frustrations, and achievements.

There is another, much earlier inspiration for the book – Charles David Keeling, or Dave, as I call him. At his dining room table, Dave consistently impressed me with his drive and passion for science and the importance of providing robust, unassailable data. I extend my deep appreciation to Ralph Keeling for sharing some of Dave's story in the foreword and showcasing how his father's Mauna Loa CO_2 record has shaped the new discipline of environmental science.

Many of the ideas and activities that I present in this book were developed over time working and teaching with colleagues at the University of British Columbia. I would particularly like to acknowledge Douw Steyn and Kai Chan who helped me develop approaches for reading scientific papers and writing scientific proposals that I have included in this book. Valentina Radic, Phil Austin, and three anonymous reviewers provided invaluable critiques and suggestions that improved the data analysis sections of the book. I also extend heartfelt thanks to Sara Harris, who provided me with daily writing support, and loving gratitude to Paul Keeling, my husband, who helped with copy-editing.

It was a pleasure to work with the editorial staff at Cambridge University Press – Emma Kiddle, Sarah Lambert, and Esther Migueliz Obanos. Thank you all.

Finally, I would like to dedicate this book to my parents, Liz and Bob Ivanochko, who taught me to value intellectual curiosity, creativity, initiative, and determination.

Introduction

This book is intended to help students shift from being passive consumers of scientific content to active participants in the process of science. This transition, from the student in the classroom to the effective practitioner, can be frustrating at first. A genuine scientific question can be asked and answered in a variety of ways. There is no one correct way to tackle a problem and no one correct approach to answer a question. Starting to actually do real science can be intimidating. What constitutes a good question? How do you know what data you need to answer it? How do you convince others that what you are doing is worthwhile?

Outside the classroom, scientists do not know the answers to questions they ask in advance. Each new scientific project has its own particular sets of challenges. Therefore, no step-by-step instructional guide can provide the one correct way to tackle all problems, frame all questions, or analyze all data sets. It is possible, however, to develop a systematic approach to asking and answering scientific questions through practice:

- General questions must become focused.
- The data used must represent the scale of the processes or changes of interest.
- Data analysis should help showcase emergent patterns, cycles, events, or changes.
- The conclusions drawn must be supported by evidence.

Doing science is more than doing a calculation. Practising authentic science in a classroom requires students to ask focused scientific questions, logically develop an approach, and use real data to answer the question. Focusing a question, aligning data with the question, analyzing and presenting the results to highlight key findings, and drawing supported conclusions all require a combination of logic and creativity. When students engage in authentic science activities, they have to grapple with the uncertainly of open-ended projects. This book aims to support students through this process using a real data set as a practical example of how to iteratively ask and answer scientific questions.

Environmental science has its own unique particularities. Questions in environmental science commonly investigate an intersection between natural processes and human activities. These questions can investigate changes in space or time, on local to global scales, over minutes to millenia (or longer). In many cases, environmental data sets are composites of more than one signal with more than one scale of variability. To a beginning scientist, this can be daunting. Working with environmental data sets requires imagination, organization, and patience. As the logistics of working with large and complex data sets are normalized, more opportunities arise to ask and answer more interesting questions.

This book is loosely organized into three parts:

Part I Thinking Environmental Science

Chapters 1–4 are focused on developing scientific thinking skills and providing context for environmental science questions. In Chapters 1 and 2, the process of doing science is differentiated from the practice of communicating science. This distinction is important. Written scientific communication is linear with distinct headings (introduction, methods, results, discussion, conclusions) that help communicate information clearly in written form. But the process of doing science is more fluid and iterative than is depicted by a scientific publication. The misconception that the process of science is linear can be a pitfall for the beginner scientist. The inevitable iterative process of asking a question, investigating the question, gaining more knowledge, revising the question, and so on can feel unproductive.

With the development of scientific thinking, students gain the ability to read published papers and interpret scientific graphs critically. Reading a paper to find the science is different from reading a paper to find the answer. Once the process of science is illuminated within the publication, it is easier to critically evaluate what has been done to date and find knowledge gaps. At this stage, the process of asking a question, investigating the question, gaining more knowledge, revising the question, and so on can feel productive.

Chapters 2 and 3 remind us that environmental science is place based. Investigating the natural world requires us to consider the natural processes at work and the impacts that these processes have on the systems or phenomena we are interested in. Environmental data is collected by many institutions for many reasons. The extent of environmental data that is now publicly available is a resource that should not be overlooked when planning a project.

Part II Doing Environmental Science

Doing environmental science starts with recognizing that data is information. With that in mind, the focus of this book is on how to isolate particular information from a particular data set to help answer a particular question. To demonstrate this idea, one data set is used across Chapters 5–12. Employing a question-driven approach, the iconic record of atmospheric CO_2 concentrations (originally collected at Mauna Loa Observatory by Charles David Keeling and known colloquially as the Keeling Curve) is characterized statistically, decomposed, correlated, and modelled. Both general concepts and specific examples of basic data analysis are presented in each chapter. Answering one question provides new information and therefore opportunities for new questions. In this way, basic timeseries data analysis and the process of science are presented.

Students are encouraged to obtain the Mauna Loa CO_2 data set and work with it alongside reading this book. A conceptual understanding of averaging, variance, standard deviation, correlation, and regression is emphasized in the text. Practical knowledge can only be gained by doing. All calculations are explained and demonstrated in simple spreadsheets to encourage students to execute the calculations and not solely rely on software functions.

Part III Communicating Environmental Science

Writing a research proposal is an important practical step in the process of science. A successful research proposal can realize access to a research program or funding for a research project. A research proposal is also an excellent way to consolidate ideas. Chapter 13 outlines how to scope a research project, frame a research question, and ensure that the project relevance and implications are differentiated and clearly articulated. Chapter 14 tackles the abstract, the scientific summary. The epilogue synthesizes all of the information in the book by suggesting a rubric that can be used both a tool to guide proposal writing and as a tool for assessment.

In advance of each chapter, readers are prompted to think about a few ideas related to the upcoming material. These opportunities allow readers to determine their own level of comfort and competency with the material. Considering a topic in advance of the chapter allows readers to identify their own areas of confusion. Identifying personal gaps or uncertainty prepares readers to make better use of the chapter materials. I encourage readers to take the time to engage in the chapter preparation activities before reading the chapter.

Throughout the book, two themes are reiterated:

Logic: The alignment of research question, methods, results, and conclusions are discussed as part of the development of scientific thinking. This same alignment is necessary when doing data analysis, presenting results, and writing a research proposal.

Scale dependence: As environmental processes can act on multiple temporal and spatial scales at the same time, the importance of matching the scale of the phenomenon of interest with a process acting on that same scale is highlighted. It is important to actively assess scale dependence when interpreting published results, when collecting or finding data, when analyzing data, and when proposing new work.

This book will guide students to *think* scientifically, *do* basic data processing, and *communicate* a scientific idea as a proposal. Although the practical examples and data sets used in this book derive from the environmental sciences, the thinking, basic data processing, and communication skills that will be gained are applicable beyond environmental sciences.

PART I Thinking Environmental Science

Chapter 1

Preparation

1. Before reading Chapter 1, consider the following questions:
   - Where in a scientific paper are the research questions presented?
   - Where is the motivation discussed?
   - How do the authors convince you that their methods are appropriate?
2. Read the following abstract from Halpern et al. (2008).

A GLOBAL MAP OF HUMAN IMPACT ON
MARINE ECOSYSTEMS

> The management and conservation of the world's oceans require synthesis of spatial data on the distribution and intensity of human activities and the overlap of their impacts on marine ecosystems. We developed an ecosystem-specific, multiscale spatial model to synthesize 17 global data sets of anthropogenic drivers of ecological change for 20 marine ecosystems. Our analysis indicates that no area is unaffected by human influence and that a large fraction (41%) is strongly affected by multiple drivers. However, large areas of relatively little human impact remain, particularly near the poles. The analytical process and resulting maps provide flexible tools for regional and global efforts to allocate conservation resources; to implement ecosystem-based management; and to inform marine spatial planning, education, and basic research.

In point form, use words or phrases from the abstract to describe the following:

1. The motivation for the study:
   - o

2. The research objective:
   - o
3. The approach or methods employed:
   - o
   - o
   - o
4. The key results of the study:
   - o
   - o
   - o
5. The impacts or implications of the study:
   - o
   - o
   - o

1 Reading Papers to Find the Science, Not the Answer

1.1 SCIENTIFIC PUBLICATIONS

The way scientists work is not linear. A scientist does not think quietly to herself "I am following the scientific method" as she observes, hypothesizes, tests, and concludes. In fact, the process of science is much more iterative, circular, and creative than is implied by a linear model of the scientific method.

However, the way scientists communicate their work *is* linear. To effectively communicate science, it is necessary to outline how everything – the question, methodology, data, and analysis – lines up to support the conclusions that are drawn. A scientific journal article should not be interpreted as a general description of the process of science. Instead, scientific writing should be understood as one particular story relating one particular data set to one particular interpretation.

Individual scientists share their work with the broader community in the form of papers (also called articles) published in scientific journals. Each published paper is expected to add new information or a new interpretation to the existing body of scientific knowledge. Before being published, these papers are subject to a rigorous process of peer review (see Box 1.1). Once accepted by peers and the editorial staff of journals, peer-reviewed papers add to the official body of scientific knowledge. In this way, each new paper builds on what has been published before.

Scientific papers tend to follow a structure that uses headings to guide the reader through the different sections of the paper. Typical headings include *Introduction, Methods, Results, Discussion,* and

BOX 1.1 **Peer review**

Before an article is published in a scientific journal, it is reviewed both by the journal's editorial team and by experts in the field that are external to the journal. External reviewers are selected from the community of scientists who have previously published on a closely related topic. Usually, two to three external reviewers participate in the peer-review process for each paper submitted for publication.

The peer-review process is intended to ensure that the information presented in an article is logical and adds new information to the current body of scientific knowledge. Peer reviewers summarize the main contribution of the submitted article, point out perceived gaps in logic, or highlight missing, misinterpreted, or irrelevant information. The peer reviewers confirm or deny that the authors have used appropriate methods and that the evidence presented supports the conclusions drawn. The reviewers are tasked with recommending or rejecting the paper for publication, though the journal editorial team has the final say.

If the paper is accepted with revisions, the journal editor will ask the submitting author to edit the article in response to the reviews. It is important for an author to see how other experts in the field perceive the work. Sometimes, it is only the writing or organization of ideas that has to be improved to clarify the argument. Sometimes, additional results must be included to justify the interpretation and conclusions presented. When the revised paper is resubmitted, the author also has to include a response to the reviews indicating and justifying the changes made or not made.

Conclusions. This structure helps convey critical information to the reader:

- the goal of the paper, usually a research question;
- the reasons why the question should be answered now;
- the approach taken to reach the goal or answer the question;
- the reasons why this approach is sound;

- the limitations or caveats with the methods;
- the data or results generated;
- the authors' interpretation of the data; and
- the authors' answer to the question motivating the work.

At the very beginning of a scientific paper is the abstract. The abstract summarizes the research question, methods, results, and conclusions in a short paragraph. The abstract prepares readers for the logical argument that will be made in the paper. The abstract, of course, cannot provide the nuances of the full argument and therefore cannot substitute for the full paper, but it should give enough information for readers to decide whether or not it is relevant to the their needs and purposes.

It is common for students to read scientific papers to find information or evidence: generally, students focus on the conclusions. In this way, the paper is used as a source of facts or knowledge. In contrast, scientists typically read papers to evaluate the process and logic presented by the authors, just like a peer reviewer. Scientists look for gaps in logic or for evidence that could drive a new question. This is a more critical way to read a scientific paper.

Ideally, each new paper will build on what has been published before in the relevant subject area either by thinking about existing data in a new way or by presenting new data. Critical readers will evaluate the logic of the new argument or the reliability of the new data. Data, depending on how it is collected and processed, can be used to answer different questions. Critical readers will ask, "Have the authors collected and processed the data in a logical way to support their conclusions?"

1.2 DECONSTRUCTING PEER-REVIEWED PAPERS

Before you start reading a scientific article, you should know your motivation or purpose. Why are you reading this paper? Where do you look first to find the information that you want? Where is the information that you need to meet your purpose?

1.2.1 The Research Question

The title is the first interaction that readers have with an article. The reader will make a quick decision to read the paper or not based on the title. It is, therefore, very important to word the title in a way that catches the readers' attention.

Though shocking or funny titles might attract some readers, usually scientific readers are looking for specific information. Therefore, most scientific articles have very informative titles; the more specific, the better. The title of a scientific article usually conveys the research question or goal, even if it is not written as a question. When reading a scientific paper, consider rephrasing the title as a question to clarify the main goal of the work.

In addition to the title, the research question is likely to be restated at least two more times in the article: once in the abstract, once towards the end of the introduction, and, possibly, a final time in the conclusion of the article. In many cases, the research question is not actually stated as a question, but rather as a goal. Try to find these "goal" statements and rephrase them as questions to understand the actual contribution that the article is adding to the scientific literature.

Keep track of the research question so you can refer back to it as you move through the rest of the paper.

1.2.2 Relevance

It is important for science to be contextualized. In the introduction section of an article, the authors will tell readers both the "big picture" field of study and the specific details of the new work. The authors will justify the importance of this new work in two ways:

(1) By citing previously published papers to clearly outline the scientific base that the new paper is building upon.
(2) By pointing out gaps in our current knowledge or the lack of published articles on a particular topic or idea.

The introduction section of the article sets the stage for the new science that the authors intend to present.

When reading the introduction, look for sentences that explain the motivation of the work. Commonly, there is a gap of knowledge identified. The gap will be "filled" with the new information provided by the article. Therefore, words like "new" and "novel" are big clues to the specific contribution that the authors are making. "New" and "novel" indicate that the authors have access to data or technology that no one else has seen or used yet. Or, the authors might be processing or thinking about data in a fresh way. The newness of the information or approach makes the science timely and, therefore, important to share. Clearly communicating the timeliness, or relevance, of the new information helps the authors to highlight the importance of their contribution.

1.2.3 Methods

Science must be reproducible; therefore, in scientific articles, a lot of emphasis is placed on the approach taken and methods followed to answer the research question. The approach, or the path chosen by an investigator, describes the big picture decision-making that leads to the final choice of methods used. One could approach a problem using field-based techniques or lab-based analytical measurements or numerical models. The methods section of a paper explains the approach taken and describes the important steps conducted by the researcher to execute the approach. The methods section of an article is written to show that either the authors are appropriately following a well-known and vetted methodology or that their new methodology is robust and reliable.

If a methodology is well known, the limitations of the methodology are likely also well known. In order to demonstrate that the methodology is appropriate, the authors make explicit any caveats about the methods' limitations and discuss how the limitations are met. Detailed descriptions of the experimental set-up, the field location, the resolution of the data, the environmental conditions, the use of controls, etc. not only ensure reproducibility but also inform readers that the limitations of the methodology are not exceeded.

If using a new methodology, the authors will have to explain why it is reasonable. They will indicate how the new method was tested to ensure that it is reliable and outline any limitations to the new method.

When you are reading the methods section, look for descriptions that can inform your interpretation of the results:

- Are the authors employing a specific operational definition, a unique definition created specifically for use within the bounds of the study to limit or restrict the interpretation of the results?
- Are the authors making direct measurements of a phenomenon or are they using indicators or indirect measurements that have inherent limitations?
- Is there a scientific bias (see Box 1.2) discussed or identified that limits the interpretation of the data in any way?

Effective methods will produce data that can answer the research question. Make a note of the data generated and then reread the research question: Will the data generated by the study actually be useful in answering the research question? Has anything been missed?

1.2.4 Results

The figures or tables presented in the article are the best place to look for the results. The figures, with assistance from the figure caption, should be able to stand alone from the article. Critical readers will stop to look at the figures and generate their own interpretation of the results before reading on. If the figures are well done, the readers and the authors are more likely to agree on the conclusions drawn.

Effective figures will be presented in such a way that they can provide answers to the questions that readers might have:

- Where was this study performed (in a lab, in the field, on a computer, etc.)?
- What was done (an observational study, an experiment, a model, etc.)?
- How many times (number of replicates, treatments, locations, etc.)?
- Over what timeframe?

- What change was documented?
- What is the uncertainty of the results?

Graphs are designed so that the controlled variable (x), called the independent variable, is plotted on the horizontal axis, the x-axis.

BOX 1.2 **Scientific bias**

In science the word *bias* has a usage different from the public realm. In common usage, *bias* indicates that information is being presented with an opinion or personal slant. If someone is said "to be biased," they were previously disposed towards one side of an argument, possibly for personal, value-laden reasons. In contrast, the use of the word *bias* in science is not value laden. In science, *bias* is used to indicate the inherent limitations of the particular data or measurements used in the study.

The limitations of data might come from inherent properties of the measurements made or from the approach taken by the authors. For example, in 2014, S. E. Porter and E. Mosley-Thompson wanted to know if snow samples taken from the West Central Greenland ice sheet were appropriate to use to reconstruct past winter atmospheric circulation patterns over the North Atlantic. They wanted to know if the snow at a site called Crawford Point accumulated more in summer or winter. If snow at this site accumulated faster during the summer season than the winter season, i.e., the snowfall records are biased towards the summer, then the snow accumulation data from Crawford Point would not be appropriate to use for a reconstruction of winter circulation (Porter and Mosley-Thompson, 2014).

Whether or not snow accumulation at Crawford Point occurs primarily in summer or winter is not a value judgement. Articulating a bias in data, if it exists, is just another way of accurately representing the data. The evaluation of data bias does not indicate that the data is inherently substandard or inaccurate. A clear understanding of the bias present in the data actually helps researchers interpret the information provided by the data appropriately.

The dependent variable (*y*) changes in response to changes in the independent variable. The dependent variable is plotted on the vertical axis (the *y*-axis). If the data is presented as a map, you can think about the independent variable as being physical space.

Whether you are looking at a graph or a map, you should be able to say "The authors are investigating how *y* changes in response to a change in *x*."

If data has been processed in some way, the original data should be evident so readers can see the impact of the data processing. Seeing the raw data and the processed data together provides confidence that the results presented are appropriate.

When reading a paper, try to take notes directly from the figures. Write down answers to the questions listed above using as much specific information as you can. Use the graph axes to help you quantify the details: timeframe, spatial scale, treatments, and changes seen. If possible, quantify any evident changes using both the range and the magnitude of the change; both pieces of information might be useful to you.

Environmental data regularly exhibits patterns that repeat over time. Daily, seasonal, annual, sub-decadal, and multidecadal cycles are common in environmental data. Weekly patterns might be present in human-influenced data. Sockeye salmon, who migrate from fresh water to salt water and back, have a strong four-year life cycle.

Over time, the amplitude of patterns or cycles can change. Check to see if the minimum values of a cycle are consistent, always returning to the same value, and if the maximum values are consistent. Changes in the minimum, or baseline, values are as important to note as changes in the maximum values and are sometimes overlooked. Sustained, systemic change in one direction, positive or negative, is called a trend. Is it common for both a trend and multiple cycles to be superimposed over one another in environmental data?

When considering spatial data, try to identify locations that have changed the most, locations that have changed the least, and locations that have changed in opposite directions. On maps, the dependent variable is often indicated with a quantified colour scale bar.

Though contours with numbers might not be evident on the map, the colours presented indicate the value from the scale bar. Look for patterns in relation to geographic features on different scales – local, regional, and global. Consider using different spatial boundaries (watersheds, biozones, political, etc.) to help explain the patterns in the data.

Write your own statement that indicates how the variable y changes in relation to the variable x. Try to describe the patterns you see using timescales or categories to define the temporal and spatial scale of change. Quantify the amplitude of the cycles or the magnitude and range of the change you can identify. If multiple patterns are present, indicate which one seems to be the most important.

1.2.5 Interpretation

Using the information that you have extracted from the figures, interpret the results in relation to the original research question. Once you are satisfied that you have a good idea of what the study results show, compare your interpretation to that of the author. You will likely find the author's interpretation in the discussion section of the paper. The authors' interpretation will explain what the data demonstrates or what the data means. The interpretation will also indicate the importance of the results.

When reading a scientific paper, remember to differentiate between the concrete, specific, description of the data – what we call the results – and the interpretation of the data. Everyone looking at the data should be able to agree on the results; they might not all agree on the interpretation of the results.

1.2.6 Conclusions

With your set of notes in hand, you will now be able to make some conclusions supported by evidence. Remember the motivation of the authors. The research question that they started with now needs to be addressed. Based on the study results, try to draw one or two solid conclusions that address the original research question. Check your conclusions with those provided by the authors. *Do you agree? Did they go farther than you? Do they stretch too far from their results?*

The conclusions of a paper should be very closely tied to the specific results that were generated by the study. The results must directly inform the conclusions drawn. If the authors have made a conclusion that you did not, go back and see if the data exists to support their claim.

1.3 TAKE-HOME MESSAGES

- Though the scientific process is not linear, scientific writing is linear.
- Peer-reviewed papers encapsulate the body of scientific knowledge.
- Critical readers will evaluate the process of science communicated in a publication.
- When reading a paper, know your purpose.
- Focus your attention on the details that you need for your purpose.

REFERENCES

Halpern, B. S., S. Walbridge, K. A. Selkoe, et al. (2008). A global map of human impact on marine ecosystems. *Science*, 319 (5865): 948–952.

Porter, S. E. and E. Mosley-Thompson (2014). Exploring seasonal accumulation bias in a west central Greenland ice core with observed and reanalyzed data. *J. Glaciol.*, 60 (224): 1065–1074. DOI: 10.3189/2014JoG13J233.

Chapter 2

Preparation

1. Before reading Chapter 2, consider the following questions:
   - What role do figures play in communicating science?
   - Are graph styles chosen arbitrarily?
   - What is a box plot?
   - When communicating science, is including a good graph enough?
2. Imagine that you have found a new planet and you are able to measure the temperature of the planet at various locations on the surface of the planet: 10 in the northern half of the planet and 10 in the southern half of the planet. The northern data and the southern data show opposite patterns. You have now been asked to characterize the temperature of the planet over one year.

   Using your imaginary data,[1] sketch a graph that captures how you think you would best present the temperature data for this new planet.

[1] You can choose whatever temperatures and patterns you want for your new planet.

2 Communicating Science Visually

2.1 WHY USE A GRAPH TO COMMUNICATE SCIENCE?

Communicating science is difficult. The challenge is to effectively summarize scientific results in a clear and informative way. Science is based on collecting empirical data (usually numbers) and often uses very long lists of numbers or spreadsheets with multiple columns and rows called "tables." Though it is extremely important to make the raw data (the original lists of numbers) publicly available, raw data presented as such is not easy to absorb or understand.

Figures make the job of scientific communication a lot easier. Though tables can be useful if they are relatively small and clearly laid out, visual presentations of information are generally easier to interpret. Graphs or maps are used to present data effectively and highlight critical information. Figures allow the readers to see changes or patterns in the data or to compare multiple data sets easily. It is easier to visualize comparisons, patterns, cycles, model projections, and experimental treatments using graphs or figures than using tables. Maps or contour plots show spatial patterns very effectively. Appropriately scaled axes on graphs or well-designed colour schemes on maps can bring attention to specific details of a figure that might otherwise be overlooked.

2.2 COMMON TYPES OF GRAPHS

There are many ways to present data visually. The type of graph or image presented should be chosen to align with the type of data available and the key points that need to be conveyed.

2.2.1 Histograms

Histograms are used to show distributions. In a histogram, the continuous range of values covered by the data is divided into non-overlapping subranges, bins or buckets, on the *x*-axis, and the number of data points present in each bin is plotted as a relative frequency on the *y*-axis. As histograms show continuous data, it is possible to connect the contiguous bins using a line that passes through a point located at the top of the centre of each bin.

For example, the distribution of the average seasonal temperatures in London, England, from 1910 to 2017[1] is plotted in a histogram in Figure 2.1. To generate this plot, the temperature data was collated into 0.5°C bins for the *x*-axis. Choosing the size of the bins is up to the researcher. Different bin sizes will influence the level of detail present when the data is graphed. Using a histogram to present the temperature data highlights the shape of the temperature distribution curves for the different seasons.

2.2.2 Box Plots

Box plots allow the distributions of multiple data sets to be compared. Box plots summarize distribution data by clearly indicating the median value as well as the distribution by quartiles, or the data divided into quarters. In a box plot, the middle bold line represents the median data point (Figure 2.2a). The box above the median line shows the range of the 25% of the data that falls above the median, and the box below the median line shows the range of the 25% of the data that falls below the median. The top range bar indicates the spread of the top 25% of the data and the bottom range bar indicates the spread of the bottom 25% of the data. Given this convention, the symmetry or asymmetry, skewness, around the mean value can easily be seen in box plots. Sometimes outliers, or extreme data points, are included above or below the box plot.

[1] www.metoffice.gov.uk/climate/uk/summaries/datasets#yearOrdered, last accessed July 2018.

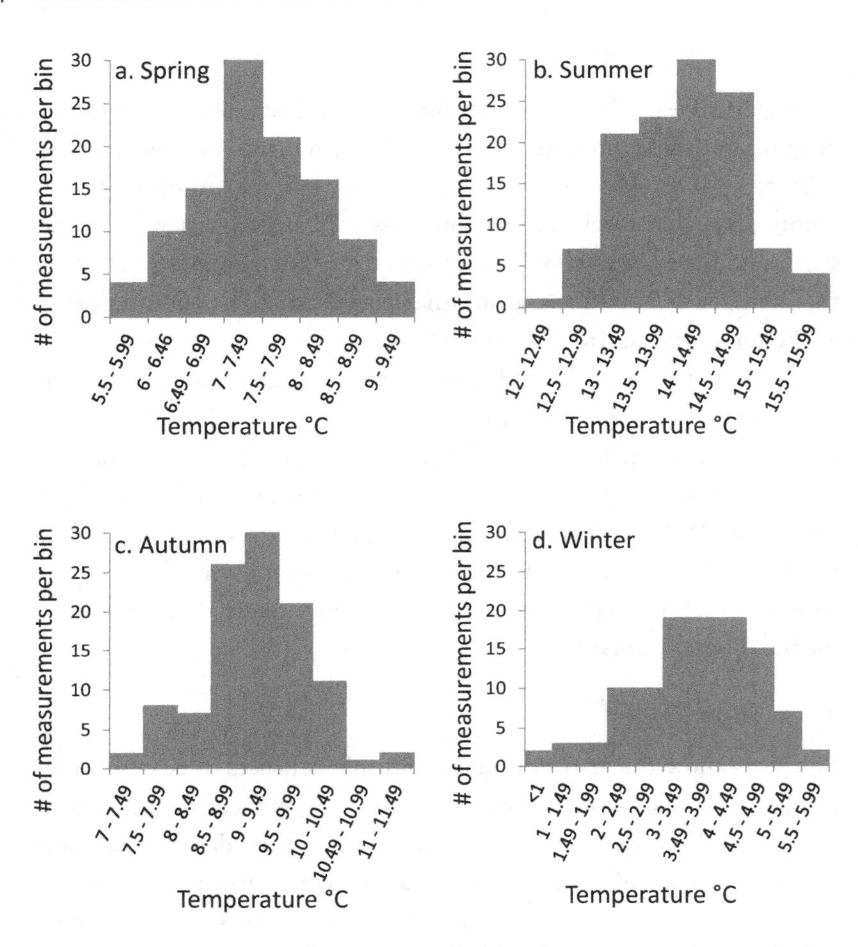

FIGURE 2.1 Seasonal temperature (°C) distributions in London, England, from 1910 to 2017. (a) Spring: March–May. (b) Summer: June–August. (c) Autumn: September–November. (d) Winter: December–February. Data from the UK Met Office, www.metoffice.gov.uk/climate/uk/summaries/datasets#yearOrdered, last accessed July 2018.

The same data[2] from Figure 2.1 is plotted again in Figure 2.2b. In this example of comparing seasonal temperature distributions from London between 1910 and 2017, the box plot can be used to

[2] www.metoffice.gov.uk/climate/uk/summaries/datasets#yearOrdered, last accessed July 2018.

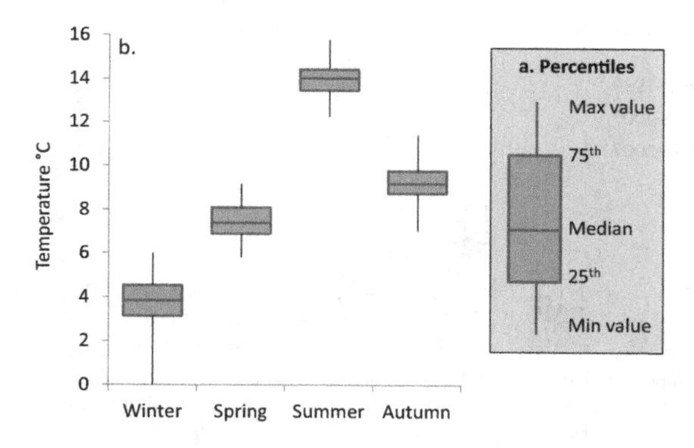

FIGURE 2.2 Seasonal temperature (°C) distributions in London, England, from 1910 to 2017. (a) Visual explanation of a box plot and percentile distributions. (b) Spring: March–May; Summer: June–August; Autumn: September–November; Winter: December–February.
Data from the UK Met Office, www.metoffice.gov.uk/climate/uk/summaries/datasets#yearOrdered, last accessed July 2018.

summarize the information very efficiently. The box plots take up less space and present all of the data on the same axis so that the differences in the temperature ranges for each season can be directly compared. The box plot does not demonstrate the shape of the distribution curve with as much detail as a histogram but it allows for an easy comparison between different distributions.

2.2.3 *Bar Graphs*

Bar graphs are used to compare values: the height of the bar represents the value of the variable. Bar graphs are regularly used to show how something has changed after an experimental treatment or from one time or place to another. Bar graphs can be stacked and colour-coded to compare how subcomponents of the value change in addition to the total value. As bar graphs do not usually represent continuous data, but rather discrete or distinct data sets, the bars of the graph are separated from one another and no line is used to connect the top of the bars.

For example, changes in the amount of carbon stored in or emitted from the world's forests (Pan et al., 2011) are data that can be presented using bar graphs (Figure 2.3). Different snapshots in time

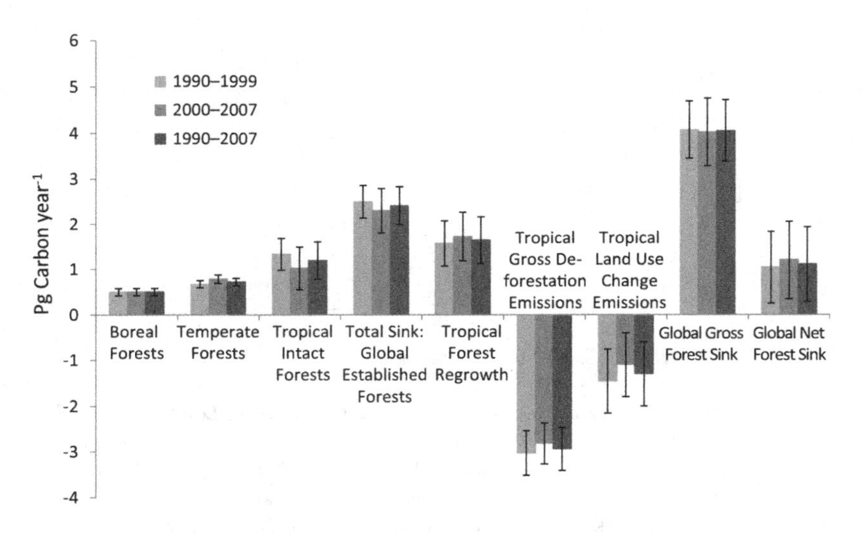

FIGURE 2.3 Comparing the amount of carbon stored in or emitted from the world's forests between two decades: 1990s and 2000s. Positive numbers indicate carbon dioxide removal from the atmosphere. Negative numbers indicate carbon dioxide supply into the atmosphere. Global Gross Forest Sink is the sum of Established Forests and Tropical Regrowth Forests. Global Net Forest Sink is the sum of Established Forests and Tropical Land Use Change.
Data from Pan et al., 2011.

at one location can be compared in serial bars, and different locations or categories can be compared as new sets of bars. The consistent colour-coding in each new set of bars allows the data to be understood easily. Error bars can be added to each data set to demonstrate the uncertainty of the measurements.

2.2.4 *Pie Charts and Donut Charts*

Pie charts and donut charts show proportions. The whole pie represents 100%. Each subsection of the pie represents a portion of the total. Pie charts are simple representations of how a whole is made up of aggregate parts. Pie charts are not commonly used in scientific analysis (they are more common in the social sciences) but sometimes they can be useful.

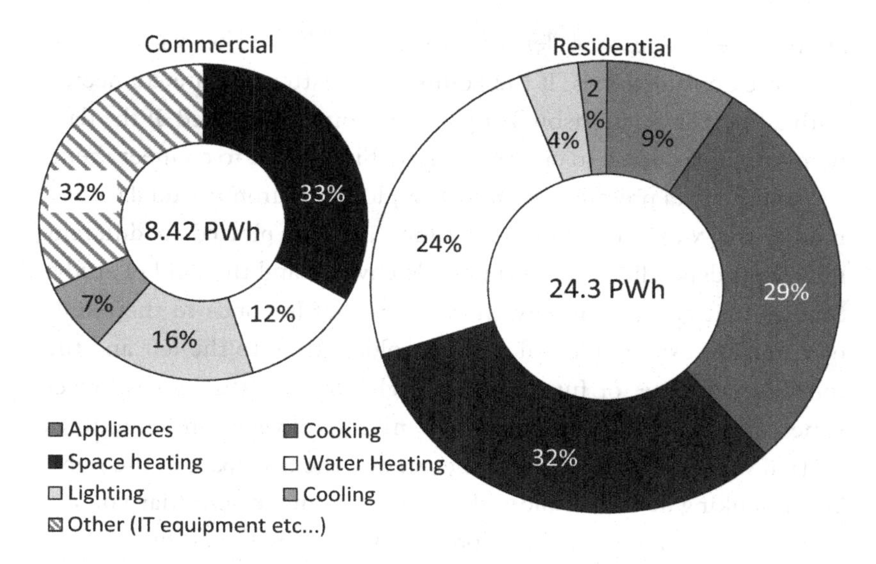

FIGURE 2.4 A comparison of global residential and commercial sector building energy consumption in 2010.
Data is from the Intergovernmental Panel on Climate Change Report 5 (Lucon et al., 2014).

For example, a comparison of residential and commercial energy consumption (Lucon et al., 2014) can be effectively presented using a pie chart (Figure 2.4). The pie chart allows you to see the total energy consumption by category within the residential and commercial sectors of society. The consistent colour-coding for all categories of use allows for a direct comparison of the proportional energy usage per sector. This type of data could also be presented using a stacked bar graph. The particular choice of presentation is up to the author.

2.2.5 *Scatter Plots*

Scatter plots are used to present two variables of the same data set. If the data is experimental data, the *x*-axis is used for the independent variable, the variable that is controlled or manipulated. The *y*-axis is used for the dependent variable, the variable that responds to change in the independent variable. If the two variables are unrelated to one another, i.e., there is no dependent variable, then either variable can be plotted on the *x*-axis. If the points on a scatter plot are connected with a line, the relationship between the points is implied. If the data points are not related to one another, no line is used to connect them.

If the data presented on a scatter plot is environmental data or a model, the *x*-axis is commonly time. Time is plotted in different directions depending on the timescale covered and the field of study. Studies that cover relatively short timescales (seconds to thousands of years) are usually plotted with the oldest time to the left and the most recent time, or future, to the right. Studies that cover longer timescales, hundreds of thousands to millions of years, are sometimes plotted with the present day to the left and the deep past to the right. When looking at scatter plots showing time, make sure that you are correctly identifying the direction of time used by the author.

When there is a lot of data presented in a time series, it is common to only show the line and omit the data points on the graph. However, this must be done with care, as it is important to clearly indicate how much data there is and how the data is distributed in time. Sometimes more than one data set is plotted on the same graph so that changes over time can be compared. In this situation, there might be two or more *y*-axes used if the scales or units of the variables cannot be represented on one axis appropriately.

A time-series data of global, Northern Hemisphere, and Southern Hemisphere annual average temperatures can be compared on a scatter plot (Figure 2.5) (GISTEMP Team, 2018; Hansen et al., 2010). To make the comparison relative, all three data sets have been transformed into a non-dimensional (unit-free) °C change from a base time period; in this case, from 1951 to 1980.

This type of transformed plot, called an anomaly time-series plot, is used to compare relative changes rather than to compare absolute values. Anomaly analysis is powerful because it allows data sets from different locations to be compared on the same scale. Rather than comparing place-specific values that might be influenced by a variety of local factors (comparing actual Antarctic temperatures with actual

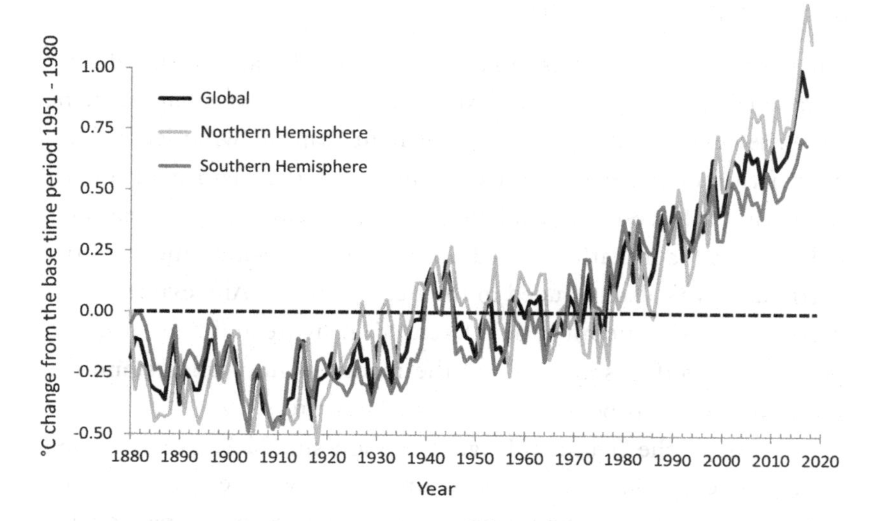

FIGURE 2.5 Global, Northern Hemisphere, and Southern Hemisphere combined land-surface air and sea-surface water temperature anomalies (°C) from 1880 to 2018. The dashed line indicates the baseline temperature calculated as the average temperature between 1951 and 1980.
Data is from GISTEMP Team (2018).

Equatorial temperatures), comparing anomalies, or non-dimensional data, allows you to focus on cycles or change evident in more than one data set (how much Antarctica has warmed since 1880 compared to how much Equatorial regions have warmed since 1880).

2.2.6 *Maps*

Maps are excellent for showing spatial variability. Multiple maps in sequence, or maps reflecting a change, can show spatial variability over time. Three-dimensional information can be presented using contour lines. Well-chosen colour schemes that identify gradients or change in space can make key features of the data stand out. However, as generating maps requires specialized software and colour images, maps will not be discussed in detail in this book.

2.3 FIGURE CAPTIONS

All figures (graphs, tables, and images) must be accompanied by a numbered figure caption: an explanation of the information being presented in the figure. Typically, the figure caption will include both a brief title of the figure that communicates the main point of the figure and a legend that explains all of the symbols, abbreviations, colours, and annotations used in the figure. Sometimes a brief methodology is also included in the figure caption. Any external data presented in a figure must be referenced. A figure and figure caption presented together should allow the reader to fully understand the information that is being presented in the figure.

Though the figure and figure caption should be able to stand alone, in scientific writing, the figure is always referred to (by the figure number) somewhere appropriate in the text of the paper. This reference links the figure to the text and allows readers to connect the information presented in the figure with the author's narrative.

2.4 DESCRIBING GRAPHS: CHANGE, PATTERNS, CYCLES, AND EVENTS

A well-designed graph with an informative caption will help you convey your key message. Choosing the right graph will allow the reader to easily see the information that you want them to see. But it is also important to be able to describe in prose, using words and numbers, the information that is presented visually.

Recall from Chapter 1 the importance of extracting information from a graph to generate your own interpretation of the results, evaluate the author's interpretation, and draw meaningful, supported conclusions. If, as an author, you leave the interpretation solely up to the reader, they might generate an interpretation of your results that is quite different from yours. The more clear and specific you can be when communicating your results, the more likely you will be able to persuade the reader that your interpretation is correct.

To explain your results in writing, pretend that you are describing your figure to an artist who has to draw it without looking at it. You can start by making a list of what the artist would need to know.

- What was done: an experiment, an observational study, or a model?
- What change is important: over time, between treatments, or between sites?
- What variables are presented: what is *x*; what is *y*?
- How does *y* change in response to a change in *x*?

From this information, the artist could decide on the type of graph to use and could label the axes, but would need more specific information to draw the following details:

- What is the scale on the *x*-axis?
- What is the scale on the *y*-axis?
- Is there a pattern to how the data changes from the start to the finish?
- Does the pattern have a cycle with an associated timescale?
- Is there more than one cycle?
- What are the frequencies and amplitudes of the cycles you see?

Using numbers, quantify the

- conditions over which the change happens,
- magnitude of the evident change,
- frequency and amplitude for all of the patterns or cycles of importance, and
- magnitude and scale of any unique event of interest in comparison to the background conditions.

If the plot is quite complicated, make sure you highlight the main result that you want your readers to notice. The take-home message from the graph or image must be clear.

2.4.1 Example 1: Describing a Bar Graph

Look at the graph again in Figure 2.3. The graph demonstrates carbon uptake and release from the world's forests from 1990 to 2007. Here is an example of a summary that captures and quantifies the main points of the graph in prose:

Boreal, temperate, and tropical forests all act as carbon sinks, removing approximately 0.5, 0.72, and 1.19 Pg of carbon, respectively, from the atmosphere every year. No significant change in the carbon uptake from these different forest categories happened from the 1990s to the 2000s. Cumulatively, the world's forests remove approximately 2.41 Pg of carbon from the atmosphere every year.

However, between 1990 and 2007, the carbon release to the atmosphere by tropical deforestation (2.94 Pg C/year) exceeded the carbon uptake from tropical reforestation (1.64 Pg C/year), resulting in a net release of 1.3 Pg of carbon from tropical forests back to the atmosphere every year.

Considering the impacts of tropical deforestation between 1990 and 2007, the net uptake of carbon from the world's forests was approximately 1.11 Pg of carbon every year.

2.4.2 *Example 2: Describing a Scatter plot*

Look again at the global, Northern Hemisphere, and Southern Hemisphere annual average temperature changes since 1880 presented in Figure 2.5. Here is an example of a summary that captures and quantifies the main points of the graph:

Global, Northern Hemisphere, and Southern Hemisphere annual average temperatures have all increased since the industrial revolution in the late 1800s. In addition to subdecadal variability, with an amplitude of approximately 0.3°C, all three records demonstrate an increasing trend since the baseline period 1951–1980. Between 1978 and 2018, the annual average Northern Hemisphere temperature has increased ~1°C (0.03°C/year) while the annual average Southern Hemisphere temperature has increased only ~0.7°C (0.18°C/year).

2.5 TAKE-HOME MESSAGES

- Communicating scientific results is difficult, but well-designed graphs make it a lot easier.
- Identifying the independent (*x*) and dependent (*y*) variables allows you to determine how *y* changes in response to *x*.

- Different types of plots are used to highlight different features of the data:
  - » Histograms show distributions.
  - » Box plots compare distributions.
  - » Bar graphs show change between two or more conditions.
  - » Pie charts show proportion of a whole.
  - » Scatter plots compare two or more variables.
  - » Maps show spatial variability.
- All figures must be accompanied by a numbered figure caption that allows the figure to be understood independently from the text.
- In the text, always include a quantitative description in prose that highlights the specific, intended take-home message of a figure.
- All figures must be referred to, using the figure number, somewhere in the text.

REFERENCES

GISTEMP Team (2018). GISS Surface Temperature Analysis (GISTEMP). NASA Goddard Institute for Space Studies. Last accessed on 17 July 2018.

Hansen, J., R. Ruedy, M. Sato, and K. Lo (2010). Global surface temperature change. *Rev. Geophys.*, 48: RG4004. DOI: 10.1029/2010RG000345.

Lucon, O., D. Ürge-Vorsatz, A. Zain Ahmed, et al. (2014). Buildings. In: O. Edenhofer, R. Pichs-Madruga, Y. Sokona, et al. (eds.), *Climate Change 2014: Mitigation of Climate Change. Contribution of Working Group III to the Fifth Assessment Report of the Intergovernmental Panel on Climate Change.* Cambridge University Press, Cambridge, UK and New York.

Pan, Y., R. A. Birdsey, J. Fang, et al. (2011). A large and persistent carbon sink in the world's forests. *Science*, 333 (6045): 988–993. www.metoffice.gov.uk/climate/uk/summaries/datasets#yearOrdered. Last accessed July 2018.

Chapter 3

Preparation

1. Before reading Chapter 3, consider the following questions:
   - What natural cycles commonly influence environmental data?
   - Why do environmental scientists focus on patterns in time and space?
   - What is the difference between a process and an event?
2. Fill in the following spreadsheet with an appropriate timescale (seconds, minutes, days, years, decades, centuries, millennia ... billions of years) and spatial scale (micrometres, metres, kilometres, 100s kilometres, 1000s kilometres ...) for each word or phrase on the list. You do not have to be exact but try to capture the scales within an order of magnitude.

	Timescale	**Spatial Scale**
Cell division		
Plate tectonics		
Population growth		
An earthquake		
Evolution		
Ocean circulation		
Weather		
Migration (pick an animal)		
Glaciations		
Soil development		

3. On the graph below, plot your timescales (x-axis) and spatial scales (y-axis) from the spreadsheet in Question 2. Label each point with the processes in Column A.

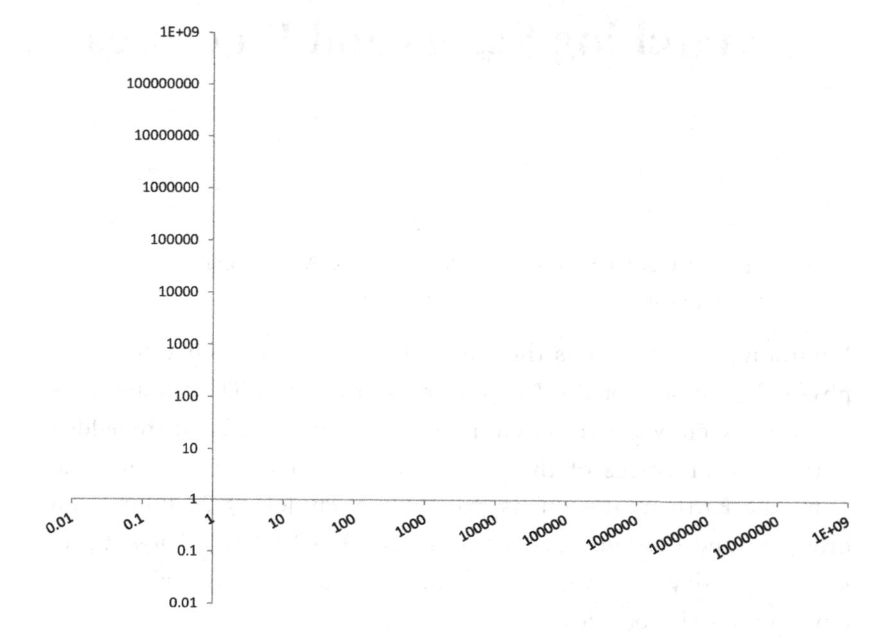

1E+09
100000000
10000000
1000000
100000
10000
1000
100
10
1
0.1
0.01
0.01
0.1
1
10
100
1000
10000
100000
1000000
10000000
100000000
1E+09

3 Matching Scales and Processes

3.1 TIMESCALES AND SPATIAL SCALES OF ENVIRONMENTAL PROCESSES

Environmental science is the study of the biological, chemical, and physical processes of the Earth, our environment. The interactions of human society and the environment are place-based and embedded within the processes of the Earth itself. The Earth is orbiting the Sun. The Earth rotates on its axis once a day (every 24 hours) and orbits around the Sun once a year (every 365.25 days). These timescales (one day, one year), and more, are inextricably linked with environmental processes.

To understand environmental phenomena, changes in environmental phenomena must be matched with the processes that influence them. Aligning timescales and spatial scales with influential processes is a fundamental aspect of environmental science.

3.1.1 Earth Basics: Cycles and Timescales

Many of Earth's processes are cyclical, repeating over and over in time. Cycles are described using the language of waves. A cycle has both a timescale – the time it takes for one cycle to be completed – and an amplitude. In a sinusoidal curve, the amplitude is the distance between the maximum value and the mean (Figure 3.1). The peak-to-peak amplitude is sometimes used when a cycle is asymmetric. The timescale of a cycle can be described in two ways:

- The wavelength (λ) of a cycle is the amount of time that it takes for the cycle, starting at one position, to return to the same position in the cycle. For example, the amount of time it takes to go from the maximum value in

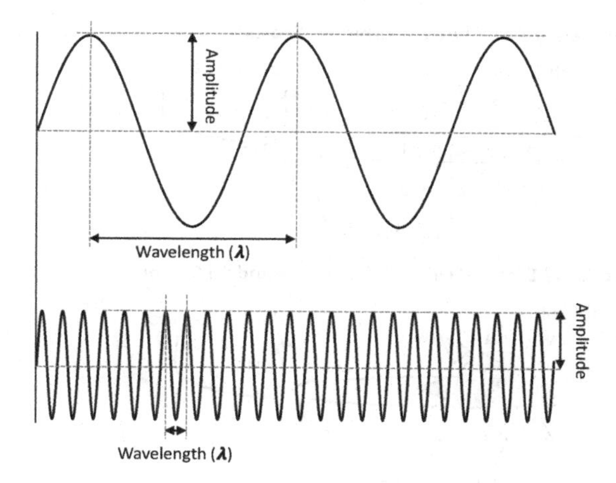

FIGURE 3.1 Describing cycles using amplitude and wavelength.

the cycle back to the next maximum value, or from the minimum value in the cycle to the next minimum value.

- The frequency is the inverse of the wavelength (λ) and is calculated as $1/\lambda$. Therefore, as the wavelength decreases, the frequency increases.

These cycles can vary on timescales (wavelengths) from hours to 100,000 years and longer depending on the process that is generating the cycle. Time series analysis is a way to identify cycles based on their frequency – the number of cycles occurring per unit of time. Most environmental data sets collected over time will reflect cycles with frequencies that are driven by the relationship between the Earth and the Sun.

There are three critical elements of the relationship between the Earth and the Sun that determine the environmental conditions, cycles, and spatial patterns that we experience on Earth:

- the angle of the Earth's axis in relation to the orbital plane,
- the direction of the axis in relation to the Sun, and
- the shape of the Earth's orbit.

If the Earth's axis was perpendicular to the plane of its orbit, all locations on the Earth would experience 12 hours of light and 12 hours

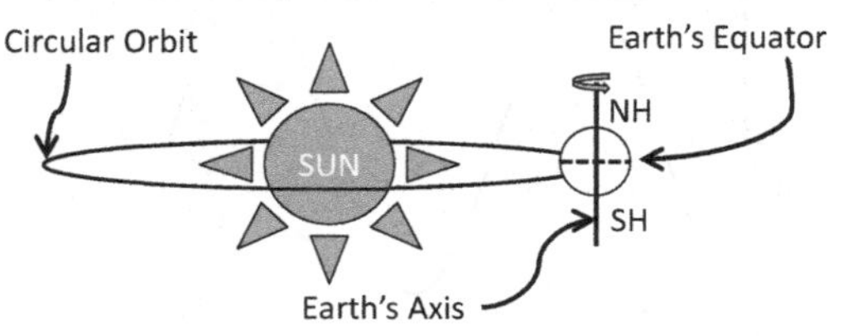

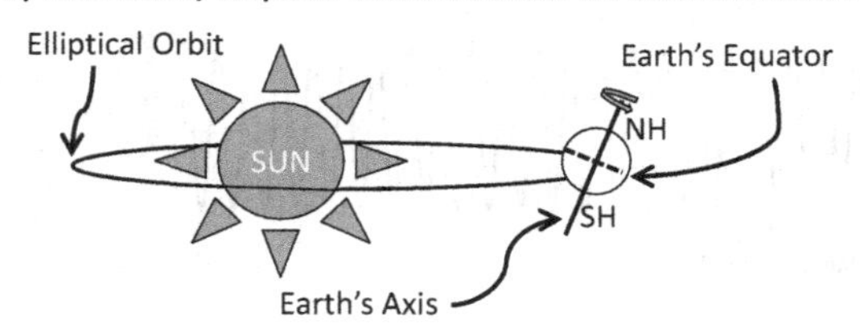

FIGURE 3.2 Comparing the impact of orbit shape and axial tilt on the Earth's seasons in the Northern and Southern Hemispheres. (a) A hypothetical situation showing a circular orbit and no axial tilt. (b) A depiction of the Earth's elliptical orbit and axial tilt (not to scale).

of darkness, with dawn and dusk as transitions. If the Earth's axis was perpendicular with respect to the plane of its orbit, and the Earth's orbit was circular, there would be no seasons and no seasonal differences between the Northern and Southern Hemispheres (Figure 3.2a).

- But the Earth's orbit is not circular: it is elliptical with the Sun at one of the focal points (Figure 3.2b). There are some times of the year when the Earth is closer to the Sun than it is during other times of the year. So there are seasons on Earth.
- Also, the axis of the Earth is not perpendicular to the plane of its orbit: it is tilted (Figure 3.2b). So the Northern Hemisphere and the Southern Hemisphere do not receive the same amount of solar radiation at all times. The seasons in the Northern Hemisphere and Southern Hemisphere are the opposite of each other: When it is summer in the Northern Hemisphere, it is winter in the Southern Hemisphere.

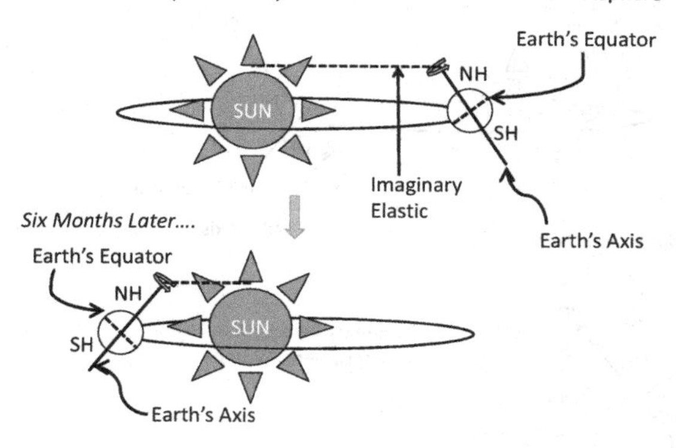

FIGURE 3.3 A hypothetical situation showing the Earth's axis pointing continually towards the Sun. This configuration would result in the Northern Hemisphere always being warmer than the Southern Hemisphere.

Currently Earth's axis is tilted about 23.4°.[1] Also, it is not tilted towards the Sun, but towards Sirius, the North Star.[2] To think about this, consider two cases:

- Imagine that there is an elastic band connecting the northern pole of the Earth's axis to the Sun (Figure 3.3). In this case, as the Earth revolves around the Sun, the northern axis (and the Northern Hemisphere) would always point towards the Sun. If this were true, the Northern Hemisphere would always be warmer than the Southern Hemisphere. This is obviously not what we observe on Earth.
- Alternatively, imagine that there is an elastic band connecting the northern pole of the Earth's axis to Sirius, a star very far away (Figure 3.4). Then imagine the Earth revolving around the Sun, but with the axis remaining connected to Sirius by the imaginary elastic band. In this case, for a portion of the year, the Northern Hemisphere would be tilted towards the Sun, generating the Northern Hemisphere

[1] The Earth's axial tilt varies between 21° and 24° with a cycle of about 41,000 years.
[2] The Earth's axis does not always point to Sirius. The precession of the equinox, a "wobble" of the Earth's axis, has a cycle of about 26,000 years.

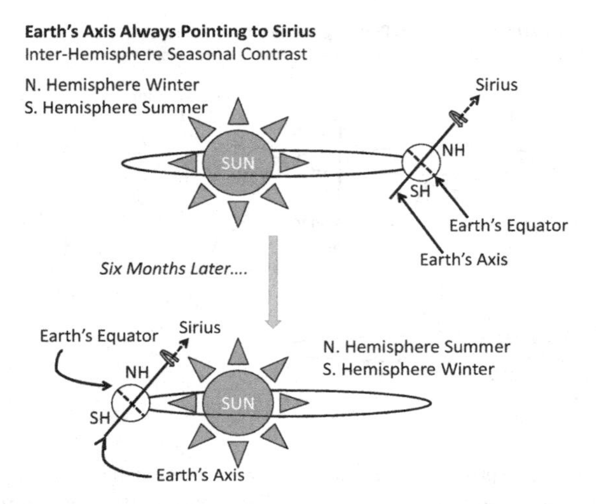

FIGURE 3.4 A depiction showing the Earth's axis pointing towards the star Sirius through the orbit around the Sun (not to scale). The position of the Earth's axis determines the seasonal contrast between the Northern and Southern Hemispheres.

> summer, and for a portion of the year, the Northern Hemisphere would be tilted away from the Sun, generating the Northern Hemisphere winter. This is indeed what we observe on Earth.

The tilt of the Earth's axis causes the differences in the lengths of the day and night that we experience, depending on our latitude. It also generates the seasonal contrast between the Northern and Southern Hemispheres. While the Northern Hemisphere is tilted towards the Sun, the days are longer and more sunlight reaches the Earth per square metre in the Northern Hemisphere: this is the Northern Hemisphere summer. It is the tilt of the Earth's axis that generates the contrasting environmental conditions between the Northern and Southern Hemispheres. It is the direction of the axis, pointing away from the Sun, that generates our interhemispheric seasonal contrast.

At one location, these different processes will influence the environmental conditions at the same time (though some processes might be more influential than others):

- Plotting hourly temperatures from Vancouver, Canada, for one month highlights the daily cycle: daytime temperatures are warmer than the night-time temperatures (Figure 3.5a). Storm tracks or weather events periodically change the amplitude of the regular day–night pattern.
- Plotting the temperature data from the same location for one year highlights a seasonal cycle with the winter temperatures, on average, colder than the summer temperatures (Figure 3.5b).
- Plotting the Vancouver temperature data for a number of years allows a repeating seasonal cycle to emerge. The growing data set shows the daily (day–night) cycle superimposed over the annual (summer–winter) cycle (Figure 3.5c).

This layering of information, short-term cycles superimposed over long-term cycles, is a common feature of environmental data.

It is fairly obvious that temperature data at a specific location will reflect the daily cycle and the seasonal cycle at that location. Other natural cycles with different longer timescales might also be evident. Anything that grows for a long period of time (trees, giant clams, coral reefs) will reflect changes in its environment on timescales longer than one year. Anything that accumulates over a long period of time (marine or lake sediments, ice in a glacier, carbonate cave deposits) will record and archive environmental changes over that time. Archived information in long-lived organisms and natural deposits provide excellent evidence for longer term natural cycles (Table 3.1).

3.1.2 Earth Basics: Patterns in Space

The Earth is a spheroid with two poles – North and South – circumscribed at its widest point by the equator. We define zones moving from the poles to the equator as polar, temperate (or mid-latitudes), and tropical. The temperate zone is sometimes divided into subpolar and subtropical. These zones have distinct features as a result of the rotation of the Earth and the distribution of solar energy around the Earth.

Over the last 4.6 billion years, heat in the Earth's interior has driven the movement of lithospheric plates on the Earth's surface, bringing the continents together and breaking them apart multiple

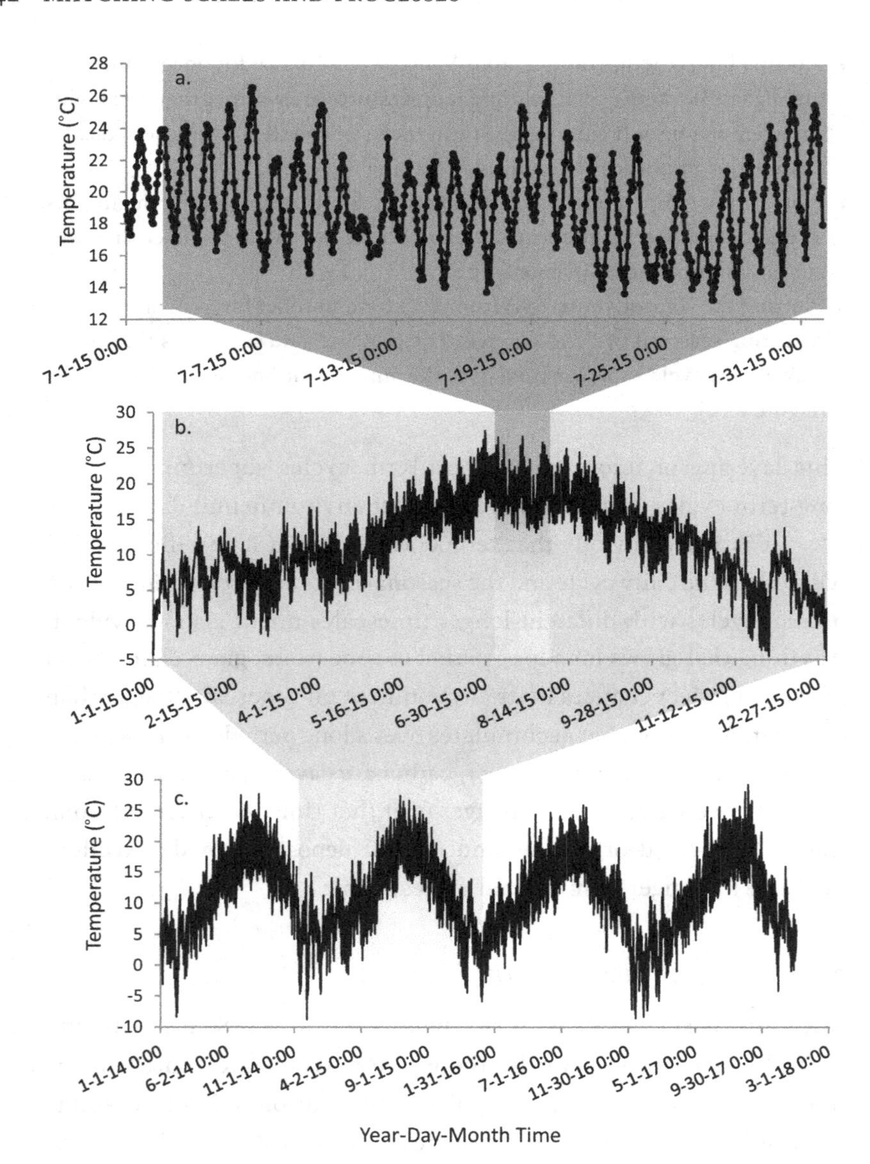

FIGURE 3.5 Hourly temperature data (°C) for Vancouver, Canada. (a) Temperature data (°C) for 1–31 July 2015. (b) Temperature data (°C) for 1 January–31 December 2015. (c) Temperature data (°C) for 1 January 2014–17 December 2017.

Data from Government of Canada Environmental and Natural Resources (2018). Historical Data, Weather, Climate and Hazard. Retrieved from, http://climate.weather.gc.ca/historical_data/search_historic_data_e.html, last accessed July 2018.

Table 3.1 *Matching environmental phenomena to timescales and environmental archives of evidence*

Timescale	Phenomenon	Examples of evidence
Diurnal	Tidal cycles	Intertidal vertical zonation
Annual	Seasonal cycle	Atmospheric CO_2 concentrations Tree rings Oyster shell rings Carbonate cave deposits (speleothems) Monsoon strength
3–5 years	El Niño–La Niña cycles (southern oscillation)	Australian drought records Peruvian fish stocks Indonesian flood records Snowpack in Western Canada
Decadal	Pacific decadal oscillation Atlantic decadal oscillation	Pacific salmon stocks Alaskan tree-ring width Temperatures in the UK Air pressure over Iceland
Millennial	Dansgaard–Oeschger cycles	Polar ice cores Marine sediments East Asian cave deposits (speleothems) Sea-surface temperature records
40,000 years	Repositioning of the intertropical convergence zone (ITCZ)	Precipitation in India Precipitation in Papua New Guinea Chinese cave deposits (speleothems)
100,000 years	Glacial–interglacial cycles	Rock deposits around the Northern Hemisphere Ice core CO_2, dust, and oxygen isotopes Marine sediments: shell compositions, sediment structure, and chemistry

times. Convergence, or the coming together, of continents forms mountain chains that slowly erode over millions of years as physical and chemical processes break them down again. This cycle has a timescale of about 500,000,000 years.

Today on Earth, we have seven continents spread out around the globe in the polar, temperate, and equatorial zones in both the Northern and Southern Hemispheres.

Polar zones have two distinct seasons: winter (dark) and summer (light). When it is winter in the northern polar region, it is summer in the southern polar region. Both poles receive very little precipitation.

The temperate zones tend to have four seasons: winter, spring, summer, and autumn (fall). The duration and intensity of these seasons will vary depending on the specific location. Precipitation is more frequent throughout the year in the subpolar zones than in the subtropical zones as a result of atmospheric circulation patterns.

The equatorial zone has two dominant seasons defined by precipitation: the wet season and the dry season. In the tropics, there is convergence of atmospheric circulation cells that results in a low-pressure, high-rainfall feature called the Intertropical Convergence Zone (ITCZ). The ITCZ shifts to the north during the Northern Hemisphere summer and to the south during the Southern Hemisphere summer.

Geographic features within these zones will define regions with distinct characteristics. Regions near oceans or large bodies of water will experience less variability in temperature and more precipitation than regions that are far from oceans. Mountains generate their own weather patterns, causing more variable weather nearby.

Just as time series analysis aligns an environmental process to a relevant timescale, geospatial analysis aligns environmental processes with a relevant spatial scale. Using geographic boundaries such as watersheds, airsheds, ecozones, or migration pathways might capture the critical features of environmental processes that would not be evident otherwise. Three-dimensional spatial analysis can highlight more complex variability associated with topography (changes in the height of land) and bathymetry (changes in the depth of the ocean), such as directional flow, gradients, or nuanced population dynamics.

Politically defined boundaries, countries, provinces, states, counties, or municipalities determine jurisdictional influence over the development and implementation of policy, legislation, and regulation. Political boundaries can, therefore, influence natural systems and processes over spatial scales that differ from geographic features. Considering natural features and politically defined boundaries together can generate insight into the layering of influences that might occur at one particular location compared to another.

3.2 PROCESSES VS EVENTS

At first, it might be difficult to align a process to a scale, temporal or spatial. It is possible that the same process could be studied on a variety of scales. In many cases, you can define a long-term process and an instantaneous (or short-term) event that are clearly related.

The number of people in a country, the population, changes as a result of a country's birth rates and death rates. To understand the long-term changes in a population, you would have to understand the factors influencing the birth rates and death rates of the country. These factors are different from the factors that would influence the birth of one particular individual. If you wanted to know why the population of a country was changing, you would not look to the birth of one person for an explanation; you would have to consider the processes that influence the population at the countrywide scale.

Average birth rates and death rates have cumulative impacts over longer amounts of time. Birth rates and death rates can also be studied regionally or by country to investigate changing global population dynamics. Information about one particular birth or death, an event, cannot be extrapolated to provide information about the collective population's change over time.

The process of mountain building is driven by plate tectonics, a global-scale process working on a timescale of millions to billions of years. On this timescale, volcanism along tectonic-plate boundaries releases gaseous CO_2, which was previously trapped in the Earth's interior, and increases the amount of CO_2 in the atmosphere. Mountain building also creates new rock surfaces (newly exposed lava beds or new mountain chains) increasing the opportunities for chemical weathering that ultimately removes CO_2 from the atmosphere. So the influence of mountain building on the concentration of CO_2 in the atmosphere has an associated timescale of millions of years.

In contrast, an active volcanic eruption occurs at one place or one region on Earth at a time and can last hours to centuries, with the average volcanic event being a number of weeks. When a volcano erupts, it sends large amounts of aerosol particles into the atmosphere. The addition of aerosols to the atmosphere generally has a cooling effect by reflecting some incoming solar radiation back to space. Locally, the aerosols can remain suspended for days to weeks before falling or being washed back to the Earth by rainfall. If the aerosols are injected high enough into the atmosphere, they can remain suspended for a few years and be circulated globally. Therefore, the cooling influence of a volcanic eruption will be most intense locally for a few weeks, but large eruptions can generate a global cooling signal lasting a few years.

The process of volcanism associated with mountain building can be differentiated from an instance, or event, of a volcanic eruption. The impacts and influences of long-term mountain building/volcanism have different associated timescales and spatial scales from the impacts and influences of a particular eruption event.

3.3 TAKE-HOME MESSAGES

- Environmental phenomena fluctuate with timescales that reflect inherent natural cycles.
- Sometimes more than one cycle of change is evident in environmental data sets.

- Considering the spatial scale of a phenomenon can add insight to an interpretation of data.
- To link an environmental phenomenon to a process, the temporal and spatial scales of change must match.

REFERENCE

Government of Canada Environmental and Natural Resources (2018). Historical Data, Weather, Climate and Hazard. Retrieved from http://climate.weather.gc .ca/historical_data/search_historic_data_e.html. Last accessed July 2018.

Chapter 4

Preparation

1. Before reading Chapter 4, consider the following questions:
   - How do we know if a storm is coming, or an earthquake?
   - How are violations of environmental regulations detected?
   - Where does the data come from for environmental indices?
   - How do you decide where and how often to collect monitoring data?
2. You have been asked to document a local race. The proposed track is below. Where would you position yourself to document the race? Why? What would you see from that position? What information about the race could you collect from your vantage point?

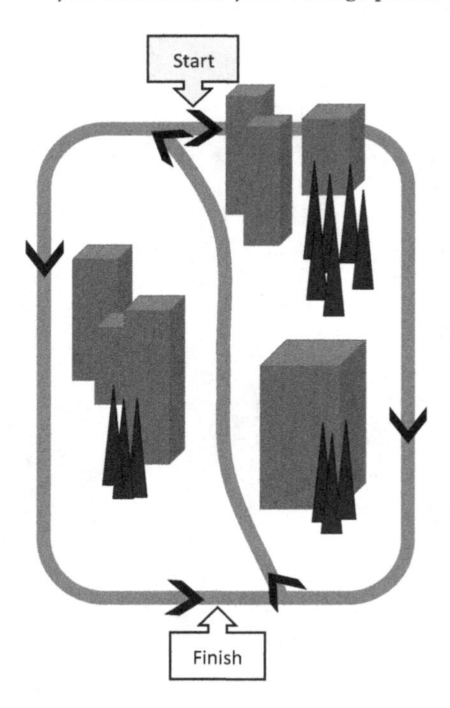

4 Collecting Environmental Data

4.1 TYPES OF DATA

Data is generated in science as the result of a designed methodology for observation. Observation can happen by watching, monitoring, testing, experimenting, or by modelling. Two types of data can be generated through observation:

- **Quantitative data**: data that provides numerical information about the phenomenon. Quantitative data sets are measurements that indicate how many, how much, how big, how small, how hot, how long, how heavy, and such.
- **Qualitative data**: data that provides verbal descriptions or non-numerical information about the phenomenon. Qualitative data provides information about things that cannot be or are not measured numerically. Examples of qualitative data could be people's opinions, experiences, and ideas, or the reasons why people act or don't act a certain way.

Sometimes qualitative data can become quantitative data. For example, a list of the birds of North America is qualitative data, but the number of each bird species present in North America is quantitative data.

The natural (or physical) sciences usually work with quantitative data to gain insight. The social sciences commonly use qualitative data, particularly about questions relating to human behaviour. Environmental science is interdisciplinary and sometimes bridges these two approaches. Depending on your research question, you might want to use both quantitative and qualitative data. The main focus of this book is on quantitative data.

4.2 WHO COLLECTS ENVIRONMENTAL DATA AND WHY?

The year 2017 was big for forest fires in British Columbia, Canada. In the summer of that year, the air quality reached 49 on a scale from 1 to 10. The forest fires started in April and a state of emergency, lasting more than nine weeks, was declared in July. A total of 1,353 fires – 552 started by people and 773 caused by lightning – burned 1,216,053 hectares of land.[1]

On the Government of British Columbia Public Safety and Emergency Services website,[2] you can easily find forest-fire statistics and geospatial data for the past 10 years, and a description of large fires in British Columbia since 1950. *Where does this information come from?*

We are all familiar with weather forecasts, which rely on environmental data collected regularly from many locations, but environmental data is also important for making sure regulations are being respected and for providing information to the public about the state of the environment. Environmental monitoring, the process of systematically collecting data to characterize environmental conditions, is done by many institutions for many reasons.

4.2.1 Forecasting

Forecasting, or predicting future conditions, relies on monitoring environmental conditions to identify patterns and deviations from the norm. Weather forecasting depends on temperature, wind speed, wind direction, humidity, precipitation, and air pressure data collected from many different locations. In coastal locations, ocean temperature data is also important. A forecast is only as good as the data on which it

[1] www2.gov.bc.ca/gov/content/safety/wildfire-status/about-bcws/wildfire-statistics/wildfire-averages

[2] www2.gov.bc.ca/gov/content/safety/wildfire-status/about-bcws/wildfire-statistics/wildfire-averages

is based. The distribution of monitoring stations and the frequency of data collection will influence the quality of the forecast.

Forecasting earthquakes or volcanic activity also requires the collection of environmental data. To identify the early warning signs of these natural hazards, regular measurements of seismic activity, ground deformation, internal ground temperature and gas discharge must be made. *But how are early warning signs recognized?*

To identify a change from the normal conditions, we need to know what "normal" is. Monitoring environmental conditions allows normal, or background, conditions to be characterized, or described statistically using means, standard deviations, maximum values, minimum values, and such, so that deviations from the background conditions can be detected.

Government agencies and public institutions commonly collect weather data and seismic data, much of which is made publicly available.

4.2.2 Ensuring Compliance of Standards, Regulations, Legislation, and International Agreements

Even if we are not trying to predict the occurrence of an event, we still might want to identify an event if it happens. Local, regional, national, and international organizations set standards or limits to protect people, property, resources, other organisms, and habitats. For example, we might want to know if nuclear testing is occurring somewhere around the world, or if there is arsenic in the drinking water, or if a population of species at risk becomes threatened, or a new invasive species is detected.

Water treatment processes, tailpipe emissions, and poaching are all monitored to ensure that legislated environmental goals and standards at regional, national, or international levels are being respected. The perceived risk associated with an event, or the importance of detecting the event quickly, determines the rigour of the associated monitoring program.

For example, the consequences of a nuclear explosion are so dire that an international treaty has been negotiated to monitor and report any evidence of nuclear activity. The Comprehensive Nuclear Test Ban Treaty[3] is not only an agreement ratified by 168 nations to prohibit the use and testing of nuclear weapons; it also includes an extensive monitoring program to ensure countries are complying with the agreement. A network of monitoring stations, the International Monitoring System (IMS), has been built to provide the data needed to detect a nuclear explosion. In 2017 the monitoring network was 90% complete. Upon completion the IMS will consist of the following:

Seismic stations: 50 primary stations and 120 auxiliary stations will monitor seismic activity to allow underground bomb detonation to be differentiated from natural seismic activity and regular mining explosions.

Hydroacoustic stations: 11 marine stations will monitor for sounds waves. Underwater explosions can be differentiated from underwater volcanism and earthquakes.

Infrasound stations: 60 atmospheric stations will monitor low frequency sound waves to differentiate between meteorites or space debris entering the Earth's atmosphere, explosive volcanoes, rocket launches, supersonic aircraft, and nuclear explosions.

Radionuclide stations: 80 air-sampling stations and 16 laboratories will detect radioactive particles and Xenon[4] released from atmospheric, underground, or underwater nuclear explosions.

As the IMS is continually monitoring seismic activity and ocean acoustics, it is also being used for early detection of tsunami waves and proving extensive amounts of data to scientists studying Earth systems.

4.2.3 Evaluating Progress towards Targets or Goals

Environmental monitoring is used to assess progress towards legislated environmental goals. The United Nations (UN) identifies major environmental threats impacting global economic and social development

[3] www.ctbto.org/ [4] Xenon is only measured at half of the air quality stations.

and facilitates international agreements to manage and mitigate these threats. Current environmental threats include global biodiversity loss (through species extinction) and climate change (driven by human CO_2 emissions). Two international agreements currently address these threats by providing frameworks for international collaboration: the Strategic Biodiversity Plan 2011–2020 and the Paris Agreement.

In 2010, the parties to the United Nations Convention on Biological Diversity adopted the Strategic Biodiversity Plan 2011–2020. One hundred and ninety-six countries are parties to this convention and 168 countries are signatories to the 2011–2020 plan. The 2011–2020 plan outlines overarching biodiversity targets, known as the Aichi Targets, and asks individual countries to provide National Biodiversity Strategies and Action Plans. The national action plans outline concrete goals that each nation feels they can meet in this time frame.

The Paris Agreement was developed in 2015 at the 21st meeting of the United Nations Climate Change Conference of the Parties (COP21) with 197 parties (countries) attending. The goal of this international negotiation was to intensify the global response to the threat of climate change, and to ultimately restrict global average temperature increases this century to below 2°C of the preindustrial average.

How can we tell if the goals are being met?

Embedded in both agreements are mechanisms for each participating country to report progress towards their goals. Participating countries submit official national targets or goals that are reviewed approximately every five years. But in order to report progress, data must be collected and evaluated to determine if, in fact, progress is happening.

National plans for limiting habitat destruction, improving protection of species at risk, and increasing marine protected areas are now common. More efforts are now being made to monitor CO_2 emissions from multiple sectors (including residential, industrial, agricultural, and transportation). Land-use changes, primarily deforestation and reforestation, which influence the CO_2 storage capabilities of land, are also being tracked. The data collected through nationally regulated monitoring programs will allow us all to evaluate if we are meeting our national biodiversity protection and CO_2 emission targets, and on track to achieve our collective international goals.

4.2.4 *Quantifying Risk for Public Information and Awareness*

In many countries, the public demands information about pollutants in the environment that pose human health or environmental risks. Indices of air quality, water quality, and ultraviolet (UV) radiation are regularly reported to the public to communicate the risks of different behaviours. Environmental indices are designed to report exposure risks, on a ranked scale, so that individuals can effectively make decisions to protect themselves.

The regular reporting of index ranks requires the regular collection of data. Environmental monitoring provides the data needed to calculate reporting metrics – based on agreed-upon quantitative assessments – of environmental risks, which allow appropriate warnings to be given. The timing of the data collection must match the reporting needs of the public. If the risks of UV exposure change hourly, then UV strength must be measured at least hourly to effectively report the risk of exposure.

Generally, local or region governments will collect data to report local environmental exposure risks. Other organizations (non-governmental organizations, the World Health Organization, the United Nations) might summarize local data sets to present national or global statistics on environmental exposure risks.

4.2.5 *Expanding Knowledge*

Environmental monitoring generates environmental data sets that cover both space and time. Scientific analysis of spatio-temporal patterns is frequently used to expand our understanding of natural processes. For example, scientists use monitoring data to study population exposure to pollutants and the consequent health effects to track pollutant transport and transformation over space and time and to model and apportion pollutant sources. Scientists monitor the environment to develop our understanding of background conditions, the present state, spatial and temporal variability, and causative relationships.

Through environmental monitoring, focused scientific efforts can evaluate the effectiveness of actions, policies, and legislation.

4.3 DIFFERENT APPROACHES TO DATA COLLECTION

As environmental processes have both temporal and spatial scales, where and when to measure environmental process need careful consideration. Both the time and spacing of measurements are referred to as the resolution of the data. Given your available resources, would it be more impactful to get high-resolution data in time (measurements made close together in time), or broad resolution in space (measurements covering a large area of ground)? This is a hard question to answer because it is difficult, time consuming and expensive, to get both simultaneously.

In 1956, Charles David Keeling was a postdoctoral fellow interested in measuring atmospheric CO_2. At the time, the role of CO_2 as a greenhouse gas that contributes to global warming was well known. But it was uncertain if the concentration of atmospheric CO_2 at any one location was dominated by local sources and local atmospheric dynamics or if a global background concentration of atmospheric CO_2 could be identified and measured accurately. Most published scientific papers at the time suggested that the concentration of atmospheric CO_2 was highly variable from place to place.

Keeling had a different idea. Based on a series of CO_2 measurements he conducted between 1953 and 1956, using a self-built pressure-based instrument to accurately measure CO_2, Keeling was confident that a consistent baseline of 310 ppm could be established (Keeling, 1998). He also noted that the variability in atmospheric CO_2 concentrations that was published in the scientific literature was likely an artifact of atmospheric mixing patterns, particularly temperature inversions that restrict atmospheric mixing and isolate the boundary layer atmosphere from the well-mixed upper atmosphere.

During 1957–1958 the measuring of atmospheric CO_2 was a priority for an international science initiative called the International Geophysical Year. There were differing ideas, however, on how the data should be collected. Keeling was convinced by his own previous measurements that atmospheric CO_2 was well mixed in the atmosphere. He was particularly interested in taking continuous measurements in one location to identify annual, sub-decadal and longer-term patterns of atmospheric CO_2. Others, who still believed that CO_2 concentrations in the atmosphere were highly variable from place to place, were more interested in a global snapshot of CO_2 that would be generated by taking measurements in as many different places as possible.

The compromise approach included both the establishment of a continuous monitoring program on Mauna Loa, an isolated volcanic mountain on the island of Hawaii, and some shorter-term initiatives: measurements were taken from the South Pole, the Southern Ocean, the Arctic Ocean, and the upper and lower atmosphere above the North Pacific Ocean. A second continuous collection site was established in La Jolla, California, next to the Scripps Institute of Oceanography, where Keeling was working.

In 1960, Keeling published results from three years of measurements demonstrating an annual cycle, driven by the net CO_2 uptake and output of photosynthesis and respiration, and evidence that the baseline value of 310 ppm was increasing annually (Keeling, 1960). Both of these results were surprising at the time. The efforts during the International Geophysical Year established the ongoing Mauna Loa CO_2 data set that is used extensively in later chapters of this textbook. The graphic representation of Mauna Loa data set is now colloquially known as the Keeling Curve.

It is worth taking a moment to consider the impact that the 1957–1958 CO_2 sampling protocol had on our understanding of

Earth's processes. The scientists involved with the International Geophysical Year were debating how to gain the best understanding of atmospheric CO_2, given the available resources. The big question was how to prioritize sampling resolution in time and space:

- Keeling's approach was to prioritize sampling in time. The thinking here was that a high-resolution record – a data set comprising hourly or daily samples collected continuously for years from one location – would provide critical details that could help us understand dynamic processes.
- The other approach prioritized space and the importance of getting a global data set – a snapshot in time – to understand regional variability and establish a global average value for atmospheric CO_2.

The high-resolution sampling from the Mauna Loa observatory has now provided a near continuous record of atmospheric CO_2 since 1958 that captures annual variability in CO_2 concentrations driven by plant photosynthesis and respiration, sub-decadal CO_2 variability driven by El Niño/La Niña cycles, and the continuous increasing contribution to atmospheric CO_2 concentrations from human burning of fossil fuels. This same record has been used to study carbon sinks, mechanisms that remove CO_2 from the atmosphere, and other aspects of the carbon cycle.

Since 1961, more atmospheric CO_2 monitoring programmes have been established. The addition of stations around the world has developed our understanding of regional and inter-hemispheric variability in both the timing and the amplitude of the atmospheric CO_2 cycles.

4.4 DIRECT AND INDIRECT MEASUREMENTS

Collecting data about current environmental conditions is relatively easy: the measurement can likely be done directly. We can measure atmospheric temperature or sea-surface temperature by placing a thermometer in the air or in the water. We can measure emissions from waste incinerators by placing probes or collection devices in the stacks. You can count turtles returning to a spawning bed. When you actually measure the variable that you are interested in, you are making a direct measurement.

But sometimes a direct measurement is not possible. We cannot get direct measurements of environmental conditions in the past. It is hard to collect data from remote locations or from populations of endangered animals. If a direct measurement is not a possibility, sometimes we can use indirect, or proxy, measurements instead.

Ice cores drilled from glaciers, sediment cores recovered from the ocean floor, and carbonate deposits collected from speleothems (cave stalagmites and stalactites) allow us to go back in time to extract information about past environments. Using known, consistent, relationships between oxygen isotope fractionation and temperature, we can use oxygen isotope ratios ($^{18}O/^{16}O$) from ice, shells, and carbonate deposits to reconstruct past temperatures. We can use tree rings to tell us about past rainfall, or pollen counts in lake sediments to tell us about what plants were present at a specific location at different times in the past. Animal scat can indirectly tell us about the location, the diet, and the health of an animal without the need to engage directly with the animal at all.

Proxy measurements rely on the assumption that the relationship between the proxy and the phenomena has not changed over time. This might not be true. Using multiple proxies of the same or a related phenomenon can help ensure that the proxy data can be robustly interpreted.

4.5 TAKE-HOME MESSAGES

- Data collected though environmental monitoring programmes can be used to predict future conditions, identify breaches of practice, quantitatively assess goals, communicate public risks, and expand knowledge of environmental systems.
- For environmental monitoring to be effective, the sampling design must capture relevant variability both in time and space.
- If the required direct measurements are not possible, proxy measurements might provide a suitable alternative.

REFERENCES

Keeling, C. D. (1960). The concentration and isotopic abundance of carbon dioxide in the atmosphere. *Tellus* 12 (2): 200–203.

Keeling, C. D. (1998). Rewards and penalties of monitoring the Earth. *Annu. Rev. Energ. Environ.*, 23: 25–82.

PART II Doing Environmental Science

Chapter 5

Preparation

1. Before reading Chapter 5, consider the following questions:
   - What comes first, the question or the research?
   - How do I write a good research question?
   - Where do I start?
2. Given your understanding of how science actually happens, draw a flow chart that represents a generalized but realistic scientific process:

5 Writing Research Questions

5.1 GETTING STARTED

A good scientific question will motivate a good research project. The process of asking questions and gaining knowledge is iterative. Asking a question directs an action: an investigation into what has been done in this field before and what is already known about this problem. Sometimes the answer to your question can be found in the published scientific literature. If that is the case, you can refine your question. Answering a question generates new knowledge, which in turn generates new questions, and so on. So your questions will become clearer and more useful as you gain information, resources, and experience in your field. Be prepared to review your research questions regularly. They may need to change over time.

Many fields of science are question driven and experimental. But environmental science is commonly exploratory. However, even for exploratory environmental science, it is necessary to articulate a concrete, clear, and focused question that can direct your data collection. Even more detailed, specific questions will arise after or in concert with the collection of data.

If the answer to your question is not known, you will be trying to answer it. As you develop your understanding of a field of study, your ability to focus on important unanswered questions will grow. Though your original motivating question might start off vague, at some point it needs to be honed to be effective.

5.2 SCOPING YOUR QUESTION

Scoping is the process of defining a study. The research question is a mini-statement of a project's scope. A well-written question will concretely tell you what exactly the study will investigate and highlight the new contribution that is being made to the body of science through the completion of the study. Below are three examples of how a research question can be improved by articulating the concrete details of the study and by changing general language to specific language.

When writing your initial research question, try to be as detailed as possible about the what, where, when, and how of your study. Try to clearly state the following:

A. What is the phenomenon you are studying?

The phenomenon can most easily be articulated as "*how Y changes in response to a change in X*." This structure sets up an opportunity to study a relationship between two or more things. The more specific you can be about the relationship you are studying, the better. For the research question to motivate action, you need to know what you will measure. Imagine plotting your results. What will be on the *x*-axis and on the *y*-axis of your central, most important graph? When you know that, you are on track to writing a good research question.

B. Where is the location of your study?

Try to accurately state the spatial scale of your project. The "location" of your study could be local: Beijing, the Weddell Ice Sheet, the alpine zone of the Spanish Sierra Nevada. It could be regional: sub-Saharan Africa, the South China Sea, the Columbia River Watershed. It could be continental: Greenland, Australia. Or it could be global. Note that a global-scale question needs global-scale data. It is extremely helpful to identify the spatial scale of your project so readers can quickly place your work in context and understand the implications of the study.

C. When is the timescale of your study?

It is important to be specific about timescales. This could be in the past: "since 2011." It could be over a specific range of time: "1952–2006." Or it could be in the future: "by 2100." It is better to identify a particular date than it is to say "in 50 years" or "over the last decade." This is because readers in the future won't understand the timescale of your study without checking your publication date. This gives readers extra work to do. An untrained reader might not notice the publication date and therefore misinterpret the information presented.

D. How will you do it?

"How" refers to the approach or methodology that will be used to answer the question. "How" could be expressed as "using marine sediments" or "geospatial analysis."

If the approach or methodology *is* the new contribution being made by the study, then the "how" will be a very important part of the question.

However, sometimes the "how" is implied or unstated if the research question is too long and cumbersome. Obviously, the "how" is a fundamental part of the science – it is the specific approach and methodology that you use to answer your question – but if the methods are not new or unique, then the how might not be included in the question.

5.3 CHOOSING YOUR WORDS WELL

Because the research question guides the science, it is important for the question to be **clear, specific,** and **concise**.

Clear language is unambiguous language: the words used are not open to interpretation or do not have more than one meaning. Unclear words can detract from the readers' understanding.

Specific language provides details. Details limit and focus the interpretation of the writing. General language can be vague and therefore not helpful in getting a particular point across.

Concise language is crisp and succinct. Extra, unnecessary words can cloud the message and bog down readers.

Complicated or big words are not a substitute for clear or specific words. In fact, complicated words can detract from clarity. Though technical language can be very specific, be careful about using technical words, or jargon, that could be unfamiliar to readers. It is generally a good idea to write a research question for an educated but not an expert audience. If you are writing a research proposal, the committee reviewing your work will likely not be experts in your specific field. Try to communicate your research intentions clearly without using jargon.

It is likely that your first try at a research question will not be your last. Looking at background information in the field can help orient you. Even if what you start with is not perfect, it is better than nothing. As your ideas develop and solidify, you can review, rewrite, and clarify your work. In the beginning, you might have a vague question. But as you gather new information and determine some limits on what you can do, you can make your question more specific.

5.3.1 Example 1: The Impacts of Climate Change on Agriculture

Climate change and agriculture are intimately linked. A myriad of questions could be asked and answered that investigate a particular relationship between climate change and agriculture. Therefore, it is necessary to indicate the exact relationship you will investigate. For example, compare the following two questions:

(a) *How has past climate change impacted crop production?*

(b) *How have changes in temperature and precipitation over the last few decades impacted wine-grape quality in Italy?*[1]

Question (a) is very general. As a reader you might have many questions. What exactly is meant by climate change? Climate change could be interpreted in a number of different ways. Would the answer to this question be different if you were studying different crops? Would it make a difference where these crops were planted? The first question leaves the reader wondering what exactly is being proposed.

Question (b) is clearer and more focused. As the reader, you know exactly what specific relationship is being studied: the relationship

[1] This example was motivated by Teslić, et al. (2015).

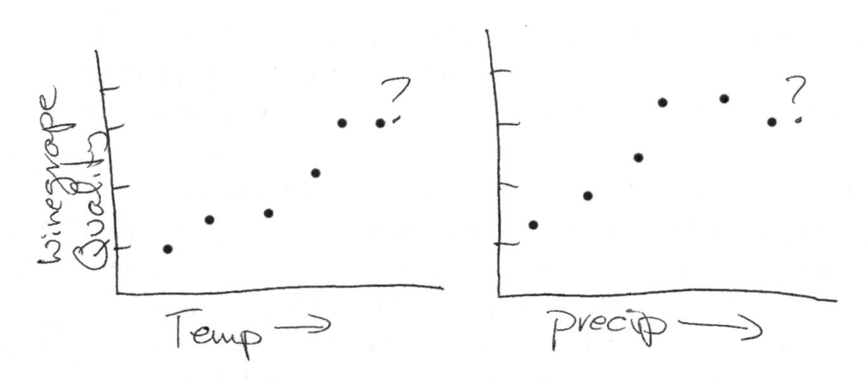

FIGURE 5.1 Conceptual sketch: How wine-grape quality changes in response to a change in temperature.

between wine-grape quality and changes in temperature and precipitation. You can imagine two graphs, each with measurable variables on the axes (Figure 5.1). The first graph would have temperature on the *x*-axis and wine-grape quality on the *y*-axis. The second graph would have precipitation on the *x*-axis and wine-grape quality on the *y*-axis.

We know where the second study will take place: Italy. The "how" part of the question is implied by stating "temperature and precipitation" and the time frame. This suggests that past temperature and precipitation data exists and that the researcher has access to that data. But the "when" part of this question, "over the last few decades," is still somewhat vague. To tighten up the timeframe, the researcher would need to know what data is actually available and decide what data to use. With a bit more investigation into available data, the "when" part of the question could be easily clarified.

The second question conveys enough information to start a project. The specific details provided in the second question help constrain the problem to wine production in Italy and direct the next steps, in this case deciding what Italian temperature and precipitation data to use. Once the data is determined, the details of the timeframe will be known and the question can be rewritten to include this information.

5.3.2 *Example 2: The Impacts of Human Activity on Marine Health*

It is relatively easy to think up interesting global-scale questions, but it is very hard to collect global-scale data. Humans do a lot of things that impact the ocean, including fishing, coastal development, and using nanoparticles in cosmetics. Scientific questions need to be answered with evidence. If you need to start with a global-scale question – that is fine. Write it down. Then you can start thinking about the data you would use to answer the question. Considering the measurements you would need to make, or the data you would need to use, will help specify your question.

Compare the following two questions:

(a) *Is human activity damaging ocean ecosystems?*

(b) *Does the presence of sunscreen in marine waters make coral reefs susceptible to viral infection?*[2]

Question (a) is vague. Reading this question, you would not be able to tell what human activity was being studied or what aspect of the ocean ecosystem was of interest. This vague question could be referring to beach construction damaging sea turtle nesting grounds or fertilizers from cornfields changing the nutrient loads of the Gulf of California. A question this vague cannot direct a science project. Where would you start?

Question (b) is very specific. "Human activity" was replaced by "sunscreen" so the reader now knows exactly what activity is being studied. The word "damage" has been replaced with "viral infection." It is now clear what aspect of the ecosystem is being studied. Question (b) will direct a scientific study on how viral infection changes as a result of changing concentrations of sunscreen. The reader can envision a graph with sunscreen on the x-axis and viral infection rates on the y-axis (Figure 5.2).

[2] This example was motivated by Danovaro et al. (2008).

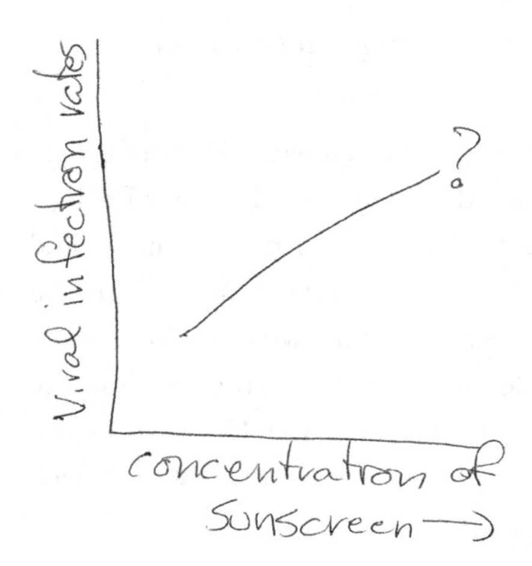

FIGURE 5.2 Conceptual sketch: How viral infection rates change in response to a change in sunscreen concentrations.

Question (b) also identifies where: coral reefs. This implies that the results are applicable to all coral reefs, not just the particular one being studied. However, if the study is not applicable to all coral species, then identifying the particular coral species that you are studying would be more informative. The "when" in the second question is the moment or duration of contact. It is the presence of sunscreen that is important to this question. The "how" is not stated. As readers, we cannot tell if this is a field study or a laboratory study. This indicates that the contribution that this study is making to the body of scientific knowledge is related more to the uniqueness of the relationship being studied than to the novelty of the methods being used.

5.3.3 Example 3: The Impacts of Air Pollution on Human Health

For readers to understand the impact of a study, the specific scientific question needs to be contextualized. Environmental science is a subset of science. Within environmental science there are other sub-disciplines: ecology, hydrology, atmospheric science, oceanography, chemistry, and more. It is helpful to consider how your question is nested within these layers of scientific inquiry. Broad, global statements help identify the broad context for the research questions. With your sub-discipline, there will be "hot topics," or active areas of research. Through reading the scientific literature in your sub-discipline (and maybe even sub-sub-discipline) you can start to identify knowledge gaps. Your specific question should direct a project that aims to fill a knowledge gap.

Compare the following two questions:

(a) *Does air pollution cause human health problems?*

(b) *What is the associated carcinogenic risk of aerosol $PM_{2.5}$ bound metals under variable weather conditions in suburban Agra, India?*[3]

Like our first two examples, question (a) is vague. There is a focus linking air pollution to human health, but is it ozone or particulate matter or something else? This question needs focus. Currently this question cannot direct action.

Question (b) answers many of our questions. "Pollution" was replaced by "aerosol $PM_{2.5}$ bound metals" and "human health problems" was replaced by "carcinogenic risk." This new language is very specific and identifies the "what" part of our question. "When" is clarified as a comparison between different weather conditions. "Where" is clearly indicated as suburban Agra, India.

Once again, the "how" is not clearly stated. It is likely that the aerosol bound metals will have to be measured directly. This is quite a unique and specific set of measurements that might not already exist. The link between metal consumption, in this case breathing in fine particles, and carcinogenic risk might be known already. The role of weather conditions and the risk potential during different weather conditions can be interpreted from the concentration data, analyzed to isolate different weather conditions, and carcinogenic potential.

[3] This example was motivated by Agarwal et al. (2017).

FIGURE 5.3 Conceptual sketch: How metal concentrations change under different weather regimes.

Table 5.1 *A summary of the changes made to clarify language in the three example questions from this chapter*

Vague/general		Clear/specific
Climate change	→	Precipitation
Past	→	Between 1952 and 2006
Crop	→	Wine-grape production
Human activity	→	Sunscreen
Damage	→	Viral infection
Ocean ecosystems	→	Coral reefs
Air pollution	→	$PM_{2.5}$

You can start to envision a comparative bar graph that identifies the type of known carcinogenic metals (on the *x*-axis) and quantifies the amounts of these metals (on the *y*-axis) present during different weather conditions (also indicated on the *x*-axis) (Figure 5.3). Exposure concentrations that breach a regulation or pose a critical health risk could be added to the graph as thresholds.

Question (b) is specific enough to direct a study.

5.3.4 Summary of Examples

The three examples above illustrate how using specific and clear language can direct a study and convey useful information. Table 5.1 summarizes the general, vague language that was replaced in these examples.

Scoping a question is part of the process of science. As you construct a question, consider how it will direct your research. Think about the specific words you chose and make sure they are leading you in the direction that you want to go.

5.4 TAKE-HOME MESSAGES

- A good research question will state the what, where, when, and how of the study.
- The research question directs the work that will be done.
- After reading a research question, you should be able to fill in specific variables for *X* and *Y* in the statement *"how Y changes in response to a change in X."*
- Write your question using clear, specific, and concise language, avoiding jargon.

REFERENCES

Agarwal, A., Mangal, A., Satsangi, A. et al. (2017). Characterization, sources and health risk analysis of $PM_{2.5}$ bound metals during foggy and non-foggy days in sub-urban atmosphere of Agra. *Atmos. Res.*, 197: 121–131. DOI: 10.1016/j.atmosres.2017.06.027.

Danovaro, R., Bongiorni, L., Corinaldesi, C. et al. (2008). Sunscreens cause coral bleaching by promoting viral infections. *Environ. Health Perspect.*, 116 (4): 441–447. DOI: 10.1289/ehp.10966.

Teslić, N., Vujadinović, M., Ruml, M. et al. (2015). Climatic shifts in high quality wine production areas, Emilia Romagna, Italy, 1961–2015. *Clim. Res.*, 73: 195–206. DOI: 10.3354/cr01468.

Chapter 6

Preparation

1. Before reading Chapter 6, consider the following questions:
   - What data could help answer your question?
   - Where can you find published data?
   - How do you know what to plot?
2. Consider the following question: How has the 2010 implementation of protected areas influenced the deforestation rates in tropical regions of Asia, Africa, and South America?[1]
   a. What data do you think you would need to answer this question?
   b. Imagining the data that you would collect, sketch a plot (or series of plots) that you think would be a reasonable way to present the data that you identified in question (a) above.

[1] Question inspired by: Blankespoor, B., Dasgupta, S. and Wheeler, D. (2017). Protected areas and deforestation: new results from high-resolution panel data. *Nat. Resour. Forum*, 41: 55–68. DOI: 10.1111/1477-8947.12118.

6 Aligning Your Question with Your Data

6.1 FINDING PUBLISHED DATA

Once you have a preliminary research question, it is time to investigate the existing data that could be used to address the question. To find existing data, start by searching the published peer-reviewed literature. Google Scholar or an open web search will not necessarily limit your search to peer-reviewed publications and can therefore be a waste of time or cause you to rely on inappropriate work, or both. The easiest way to find peer-reviewed science is to use a database at your institutional library that curates peer-reviewed publications. There are many databases that are useful. *Web of Science* is one of the most general useful databases, but you can also use databases that are more discipline specific like *Georef* or *BioMed*. Ask your librarians for advice to ensure that you are accessing the right papers for your purpose.

Once published, scientific data is usually made available for public use. Most national funding agencies now mandate that published data be publicly available and many journals stipulate as a condition of publication that the data sets they publish are accessible. Sharing data in this way recognizes the public aspect of science funding and ensures that interested parties can check the data, reproduce the published results, challenge the published interpretations, and build upon previous work.

A number of data repositories exist where the public can access published earth and environmental science data sets. Here are a few well-recognized repositories for environmental data:

- NOAA National Centers for Environmental Information (formerly the National Climatic Data Center, NCDC): climate and paleoclimate data.

- National Centre for Atmospheric Research (NCAR)
- Oak Ridge National Laboratory Distributed Active Archive Center (ORNL DAAC): biogeochemical and ecological data.
- World Data Center for Climate at DRKZ (WDCC): climate and climate-related data, specifically those resulting from climate simulations.
- NERC Data Centres: the NERC catalogue searches holdings from a network of data centres including oceanographic, freshwater, environmental, geoscience, polar, and cryosphere centres.
- Environmental Data Initiative: environmental and ecological data
- PANGAEA: georeferenced data from earth system research
- Australian Antarctic Data Centre (AADC): Antarctic data
- EarthChem: solid earth geochemistry data
- Marine Geosciences Data System: geoscience data from all oceans and continental margins
- SEANOE: marine science data
- NASA Goddard Earth Sciences Data and Information Services Center
- AEKOS TERN Ecoinformatics: Australian ecology data

If you know what data you are looking for, you can find it fairly easily. For example, if you have read a peer-reviewed paper that piqued your interest, you can search these repositories by author name, by site, or by data type, and likely find what you want. If you cannot find a published data set in a repository, you can always contact the author directly and request the data.

It is quite a bit more difficult to find data randomly. You can search these sites by subject area to see what is available. But if you want something in particular, a certain type of data from a certain location, and you haven't seen the data published, it likely doesn't exist. So when looking for data, the best thing to do is search the peer-reviewed literature for papers, identify existing data sets, and then go looking for them specifically.

If the data you need is part of routine government monitoring, it might not be present in the published literature or in these repositories of published data. Many countries allow public access to environmental data through government websites, though you might need to register to get it and in some cases there might be a fee.

When you do find the data you want, you will likely get the data in some spreadsheet format and a description of the data, called the metadata. It is important to read and retain the metadata, as it provides critical information about the data that you will likely need later. The metadata will also provide the preferred citation for the data. You should use this citation as the data reference in your work.

6.2 ALIGNING DATA WITH YOUR QUESTION

Let's look more deeply into the three example questions from Chapter 5. What data would help us answer these questions and how could we go about getting that data? The three questions are still quite open with respect to the approach. Each question could be dealt with in different ways: using existing data, collecting new data, or using a model. The approach that is ultimately used in a study is influenced by approaches previously used to answer the same (or a similar question), the interests of the individual scientist, the resources available, and the interests of the funding agencies.

6.2.1 *Question 1:* How Have Changes in Temperature and Precipitation over the Last Few Decades Impacted Wine-Grape Quality in Italy?

To answer question 1, we would need temperature and precipitation data from wine-growing regions in Italy. We would also want to know how wine quality is determined.

The temperature and precipitation data are likely to be available from a governmental organization that has a mandate to monitor environmental conditions. It is likely that there are temperature and precipitation data from every major community in Italy. We could start by looking at the wine-producing regions of Italy and see how many weather stations exist within these regions and how long the records are at each station, i.e., how far back in time they go.

To make the final decision on what data sets we would use, we would have to optimize to get as many temperature and precipitation records as possible that have comparable timeframes. We could then make the final decision on what timeframe we can in fact consider and how many weather stations we can use in this analysis. Once we have a timeframe, we could rewrite our question to update the "when" part of the question.

We would also have to decide on a measurement, or metric, of wine-grape quality: sugar content at harvest, yield, or size? Then, we need to figure out what conditions are important for achieving high wine-grape quality. We can get some of this information from previous studies about wine-grape production. We might be able to find helpful information from a wine growers' association or similar organization.

Once we have some ideas about what might be important to wine-grape quality, we need to start working with the data. Working with data is a creative process of plotting and replotting data in different ways to focus on different information. When working with data, the investigator is looking for patterns that exist or changes that might be relevant to answer the research question.

Working with the data is like starting a puzzle. You might start by sorting the puzzle pieces by function: collecting the edge pieces. Depending on the puzzle, the leftover pieces will have to be sorted in some other way, perhaps colour or shape. As you proceed with the puzzle, you might reorganize the pieces in different ways. How you choose to organize the pieces does not change the picture or the colours of the puzzle. But how you choose to organize the pieces might make it easier or harder to complete the puzzle. Similarly, how you analyse your data might make some existing features more or less evident. Focusing on finding patterns, changes, events, or cycles can help structure your data analysis, much like sorting puzzle pieces by function or shape or colour.

For question 1, based on our metric of wine-grape quality, when working with the temperature data, we might decide to plot

- daily, monthly, and annual average temperatures,
- maximum daily temperatures,
- minimum daily temperatures,
- the number of days above a certain temperature,
- the date of last frost in spring,

- the date of first frost in fall, or
- the duration of the growing season,

When working with the precipitation data, we might decide to plot

- daily, monthly, and annual average precipitation,
- cumulative seasonal precipitation,
- the number of days receiving precipitation,
- maximum daily precipitation,
- the number of days without precipitation (consecutive, annual, seasonal),
- total precipitation through the growing season, or
- the timing of precipitation through the growing season.

Depending on the data sets we use, we could investigate variability in time or in space or both. Using spatial software, we could consider comparing two snapshots in time, either in two different panels or in one panel, as a difference, to highlight the regionality of change. Alternatively, we could plot multiple time series of data to show higher frequency variability over time in more than one location.

Teslic et al 2015 considered a similar question and analyzed daily measurements of minimum and maximum temperatures and precipitation from the from the Emilia Romagna region in Italy to demonstrate that a climatic shift had occurred from 1986 to 2015. Using this approach, Teslic et al (2015) identified a change in the functional growing season for winegrapes in this region.

Working with the data helps you understand the nuanced differences in the information embedded in the data set. One set of data can be used to parse out many different relationships. After working with the data, you will know if you have results that can answer your question and will add new information to the body of scientific knowledge. At that point you might consider revising and focusing your question even more to highlight the very specific contribution you can make.

6.2.2 *Question 2:* Does the Presence of Sunscreen in Marine Waters Make Coral Reefs Susceptible to Viral Infection?

Question 2 will not likely be answered using existing data collected through a regular monitoring program. To answer question 2, we need to obtain very specific information about how a coral reef responds to the presence of sunscreen. To obtain this data, we would need to do an experiment or collect in situ (on site) measurements.

Laboratory experiments are designed to focus on one question and manipulate the independent variable to observe changes in the dependent variable. The strength of a laboratory experiment is the ability to change one variable at a time. In this case, we could take coral samples and grow them in waters containing different concentrations of sunscreen. We would collect data over time to determine the response of the coral to the presence of sunscreen in the water.

What data would we collect? Coral health is commonly measured by the amount of symbiotic algae, called zooxanthellae, present within the coral polyps. Changes in the numbers of zooxanthellae might be measured quantitatively by taking biopsy samples over time and measuring zooxanthellae concentrations, or by distinguishing changes in the colour of the coral, as loss of zooxanthellae causes coral to lighten in colour and eventually "bleach." Colour is an example of qualitative data. To transform colour into quantitative data, we might need to develop a colour index of some kind. We would also need to measure the concentration of viruses present in and around the coral sample. To understand the exact procedure for measuring the presence and concentrations of zooxanthellae and viruses, we would need to look up peer-reviewed papers that outline appropriate methods.

If we were doing a laboratory experiment, we would need to monitor the environmental conditions important to coral health: water temperature, pH, salinity, nutrients, and light levels. For simplicity, we might try to ensure that it is only the sunscreen concentrations that are varying between treatments.

A scientific control, a non-treatment replica, is included as part of the experiment to minimize the influence of variables other than the variable of interest (the independent variable). The investigator needs to be able to describe how y changes in response to a change in x. The scientific control is set up to experience the same conditions as the experiment, but is not subject to the treatment, allowing the treatment response to be isolated. In our experiment, we would include a control that experiences all of the same conditions, the same temperature, the same pH, and salinity, but that is not exposed to the sunscreen treatment. The control will allow us to see the unique effects of the sunscreen treatment.

An alternative to laboratory experiments are in situ measurements. We could approach this question by measuring sunscreen in situ, in the actual location of the living corals, and evaluating the health of the actual coral community. The strength of in situ measurements is that the variability of the natural setting is experienced. The results of in situ measurements from a natural setting are considered to more accurately reflect what is actually happening "in the field."

To conduct in situ measurements, we would still need to monitor the environmental conditions (temperature, pH, salinity, nutrients, and light levels, etc.) even if we are not consciously changing them. Our scientific control would have to be a different location that experiences very similar conditions (also measured), but with no sunscreen present in the water.

To answer a similar question, Danovaro et al. (2008) chose a hybrid approach. They took coral samples, cleaned them, and placed them in virus-free water containing measured concentrations of sunscreen. The coral samples were then incubated in situ to experience the ambient conditions of the natural setting. The amount of coral bleaching was quantified using a comparative colour scale, and concentrations of both zooxanthellae and viruses were determined by epifluorescence microscopy.

6.2.3 *Question 3:* What Is the Associated Carcinogenic Risk of Aerosol $PM_{2.5}$ Bound Metals under Variable Weather Conditions in Suburban Agra, India?

To answer question 3, we would need to find $PM_{2.5}$ (particulate matter with a diameter of 2.5 μm or less) in and around Agra, India, over enough time to capture variations in the local weather conditions. This is quite specific information. $PM_{2.5}$ is regularly collected to monitor air quality, so $PM_{2.5}$ data might exist for this region. But measuring the *adsorbed* metals, the metals attached to the surface of the aerosol particles, is not likely done as a routine measurement.

So, a question arises – *Does metal data exist or do we need to plan a methodology to actually collect and analyse the metals adsorbed on small aerosol particles in Agra?* Answering this question would require checking the Indian air quality databases to see if the data exists. In the event that there is no such data, we might need to do our own data collection to answer our research question. Our new data might then be supplemented by existing weather data from sites in and around Agra.

The second part of the research question relates to the relationship between metals and carcinogenic risk. The word "risk" suggests that we might want to generate an index, model, or exposure ranking based on different doses, the amount of metals inhaled (or consumed), in relation to the possible exposure under different weather conditions. This implies that we should know the dosage risks of the different metals.

After a field season, now with the weather data, the $PM_{2.5}$ data and the adsorbed metals data in hand, we would again start working with the data. In this example we would be looking for patterns associated with different weather conditions to find contrasting situations in which to compare high $PM_{2.5}$ values (and associated metals) with low $PM_{2.5}$ values (and associated metals). We would likely start by identifying categories of weather conditions:

- days with no precipitation vs. days with low precipitation and days with high precipitation;
- days with no fog vs. days with fog; or
- days with high humidity vs. days with low humidity.

Binary categories are easy to define; it is easy to compare days with precipitation to days without precipitation. But if you are comparing categories like "high" and "low," you will need to define and justify the threshold values between your categories. Government-mandated environmental standards would make appropriate threshold values. It is best to use well-accepted categories if they do in fact work for your analysis. If you find a new pattern that justifies a new category, that would be a very interesting result.

After working with the data and finding some interesting contrasting patterns, we could rank, using known dosage risks, the health risks associated with exposure to the metal dosage that we measured. To answer a similar question, Agarwal et al. (2017) built a model that used toxicity and dose values for individual metals and the concentrations in the air during different weather events.

It is important to remember that the results that are published in a scientific article are not the only results that the author considered. The published results are the most interesting or novel or significant results achieved. To find these results, the investigators spent time working with the data in order to extract the relevant and important information.

6.3 TAKE-HOME MESSAGES

- Previously published data likely exists in public data repositories.
- Government monitoring data is commonly available on government websites.
- To answer your research question, you might be able to use existing data.
- Your question will determine the approach you take.
- As you find data and refine your approach, you might revise your question.

REFERENCES

Agarwal, A., Mangal, A., Satsangi, A. et al. (2017). Characterization, sources and health risk analysis of $PM_{2.5}$ bound metals during foggy and non-foggy days in sub-urban atmosphere of Agra. *Atmos. Res.*, 197: 121–131. DOI: 10.1016/j.atmosres.2017.06.027.

Danovaro, R., Bongiorni, L., Corinaldesi, C. et al. (2008). Sunscreens cause coral bleaching by promoting viral infections. *Environ. Health Perspect.*, 116 (4): 441–447. DOI: 10.1289/ehp.10966.

Teslić, N., Vujadinović, M., Ruml, M. et al. (2015). Climatic shifts in high quality wine production areas, Emilia Romagna, Italy, 1961–2015. *Clim. Res.*, 73: 195–206. DOI: 10.3354/cr01468.

Chapter 7

Preparation

1. Before reading Chapter 7, consider the following questions:
   - Where do I find a description of the published data I found?
   - When I plot the data, it doesn't look right. Why?
   - There are gaps in the data. Do I have to do anything about them?
2. Consider the following graph:

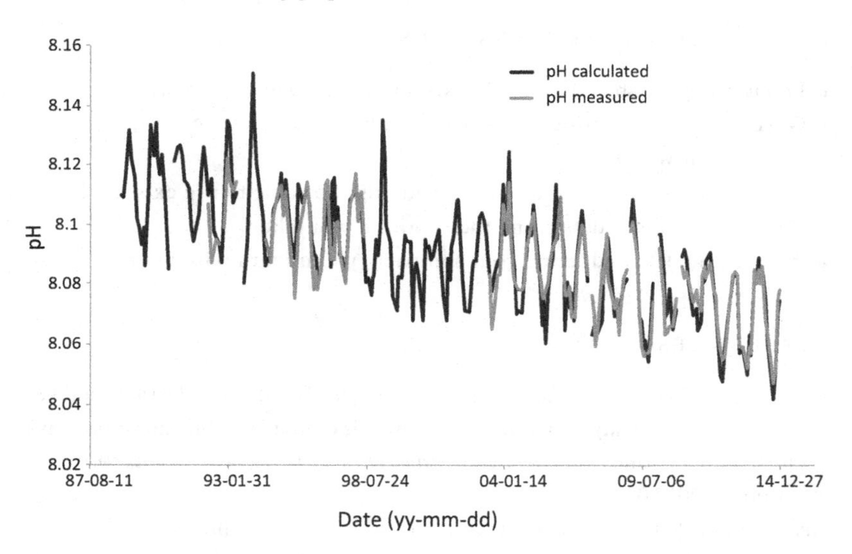

PREP FIGURE 7.1 Ocean pH data from station ALOHA in the subtropical North Pacific Ocean.

Data from the European Environmental Agency, adapted from Dore, J. E., R. Lukas, D. W. Sadler, M. J. Church, and D. M. Karl (2009). Physical and biogeochemical modulation of ocean acidification in the central North Pacific. *PNAS*, 106: 12235–12240.

a. Other than the data, try to list at least 10 of the aspects of the graph that could be changed in some way:

1.
2.
3.
4.
5.
6.
7.
8.
9.
10.

7 Working with Environmental Data

7.1 GETTING STARTED ON A PROJECT: CONTRIBUTIONS TO ATMOSPHERIC CO_2

Collecting and analyzing existing data might be a project in itself, or it might support a project that includes new data collection. In either case, it is worth practising how to find, prepare, and use existing environmental data. To start a data search, think about the research area or topic that you are interested in. It helps to pose a question.

I will start with this question:

How have concentrations of atmospheric CO_2 changed over time as a result of natural variability and human CO_2 emissions?

This question is focused enough to give me a starting point, but it doesn't have a timeframe yet. "Over time" is really too broad a statement to be useful. To be specific about the timeframe, I have to check to see what data is publicly available. Searching the *Web of Science* for peer-reviewed papers uncovers papers published using both indirect and direct data sets of atmospheric CO_2.

- Past climate (palaeoclimate) records of atmospheric CO_2 have been reconstructed from glacial ice cores around the world.
- Current atmospheric CO_2 concentrations are being monitored by different researchers, in many different locations.

There is a trade-off in these records between the timeframe covered by the data set and the resolution, or timestep interval, of the data.

Data resolution is relative. Collecting and processing data is resource intensive. Therefore, spatial data sets tend to focus either on a small amount of space in detail, capturing nuanced changes in local or regional features, or on larger amounts of space in less detail. Similarly, timeseries data sets that cover a relatively long period of time tend to provide low-resolution data. This means that the individual samples collected represent an average over a relatively large amount of time and the amount of time between samples is also relatively large. Data sets that cover shorter periods of time tend to have higher-resolution data: the samples will span a relatively short period of time and the samples will be closer together in time.

To investigate a process in time, the sample resolution must be higher than the timescale of the process of interest. Typically, monthly samples (12) are used to investigate an annual signal, daily samples (~30) are used to investigate a monthly signal and hourly (24) samples are used to investigate a daily signal. To investigate a process in space, the sample resolution must be high enough to resolve the phenomenon of interest. On a global scale, climate models typically have a resolution of 100 km^2 (at the equator) while weather forecasting uses models with 10 to 12 km^2 resolution. The spatial-scale resolution determines the timesteps required to resolve features. Coupled ocean-atmosphere models need a timestep of at least 20 minutes to resolve the interactions, while weather forecasting models use 2-minute timesteps.

7.2 CHOOSING WHAT DATA TO USE

Antarctic ice cores have provided a composite data set of reconstructed atmospheric CO_2 that spans about 800,000 years (Figure 7.1). Drilling down into glacial ice is like drilling back into time. Small bubbles trapped in the glacial ice provide samples of the past atmosphere that was isolated at the time the bubbles formed. So, past atmospheric CO_2 concentrations can be reconstructed using measurements of the CO_2 concentrations in ice bubbles.

The 800,000-year Antarctic CO_2 history has been generated from combining data from multiple ice cores from a number of sub-projects that have been published separately. The samples taken from the deepest part of the data set come from EPICA Dome C, a 3,139 m

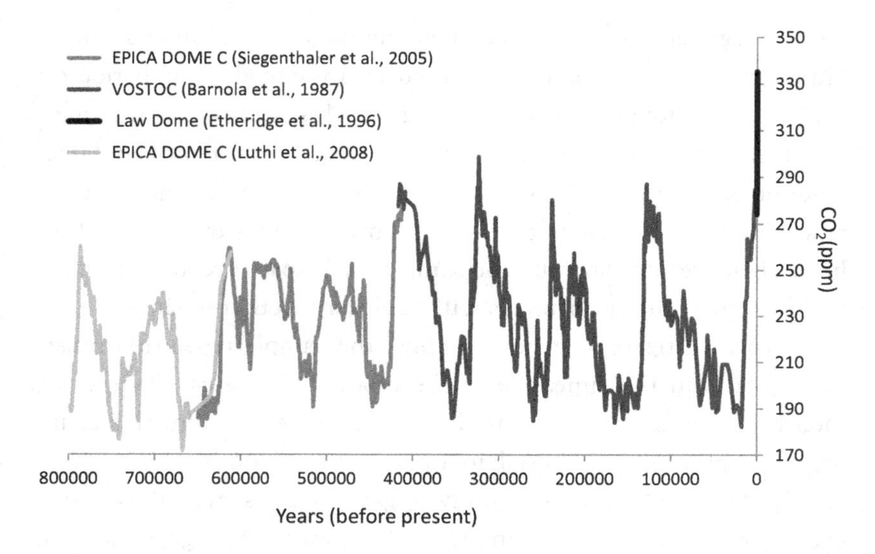

FIGURE 7.1 An 800,000-year reconstruction of atmospheric CO_2 from Antarctica. Data from four different ice cores are presented. The *x*-axis presents year before present, on the right, with the past extending to the left.

core that span about 200,000 years.[1] Above 787 m, samples were taken every 1.5 m; below 787 m, samples were taken every 5.5 m. Based on the timescale of the core, the timesteps between samples is quite variable – from about 60 years to 2,500 years. In contrast, samples contributing to the uppermost part of the compiled Antarctic data set come from a 298.055 m long ice core collected from the West Antarctic Ice Sheet (WAIS).[2] The WAIS samples span about 1,000 years and the sample resolution is approximately 10 years: a much higher resolution than the DOME C core.

Direct measurements of atmospheric CO_2 have been made at the Mauna Loa observatory in Hawaii since 1958.[3] At this site, samples are taken at least daily, so the resolution of this record is daily and even more frequent for some periods of time. The resolution of the different data sets reflects the physical conditions (amount of

[1] Bereiter et al. (2015). [2] Ahn et al., (2012). [3] Keeling et al. (2001).

ice compaction), the amount of sample needed to make a measurement, the cost of making measurements, the time needed to make measurements, and the scientific question.

My motivating question included human CO_2 emissions, so I know that I want a data set that covers the present day. To answer my research question, I will choose the high-resolution record of atmospheric CO_2 that has been collected by the Mauna Loa Observatory since 1958. This data is available from Scripps Institute of Oceanography, as part of the Scripps CO_2 Program (http://scrippsco2.ucsd.edu/data/atmospheric_co2/mlo). I recommend that you download this data and work through the examples as you read the rest of the chapter. (See Appendix A for tips on working in Excel.)

I can now revise my research question:

How have concentrations of atmospheric CO_2 changed since 1958 as a result of natural variability and human CO_2 emissions?

The Scripps CO_2 Program provides in situ CO_2 data at 10 minute, daily, weekly, and monthly resolutions.[4] The 10-minute and daily resolution records stop at 2006; weekly and monthly resolution data continue to the present. In order to decide which data set to use, I need to consider the resolution of the data that would allow me to answer my question.

My revised research question is focused on differentiating between human and natural influences on atmospheric CO_2 concentrations. Human emissions of CO_2 have increased progressively since the industrial revolution. To investigate this increase in detail, I would need a data set that has at least decadal resolution. The dominant natural cycle in the CO_2 record is the annual cycle of photosynthesis and respiration. To investigate an annual cycle, I need a data set with at least monthly resolution. There are other natural cycles, such as El Niño Southern Oscillation (ENSO) cycles, that influence atmospheric CO_2 that have sub-decadal variability. A data set with annual or

[4] http://scrippsco2.ucsd.edu/data/atmospheric_co2/mlo

monthly resolution would allow me to investigate these cycles. Therefore, the monthly data starting in 1958 and continuing to the present day will allow me to see the CO_2 variability on all of these timescales and allow me to answer my question.

If I were interested in nuanced CO_2 variability within a season, then I would choose either the weekly or daily data. If I were interested in daily cycles, I would choose the 10-minute resolution. To be able to answer a question, the resolution of the data has to be smaller than the variability that you are interested in investigating.

Now that I have my data, what do I do with it?

7.3 STEP ONE: UNDERSTANDING THE DATA

The first thing to do when you download someone else's data is to read the **metadata**, sometimes provided as text at the top of a spreadsheet and sometimes included as an accompanying README file. The metadata provides information that the authors think is important. If the authors think it is important, it is probably worth looking at.

The metadata[5] for the monthly averaged atmospheric CO_2 data set (the first five rows shown in Table 7.1) reads:

> The data file below contains 10 columns. Columns 1–4 give the dates in several redundant formats. Column 5 below gives monthly Mauna Loa CO_2 concentrations in micro-mol CO_2 per mole (ppm), reported on the 2008A SIO manometric mole fraction scale. This is the standard version of the data most often sought. The monthly values have been adjusted to 24:00 hours on the 15th of each month. Column 6 gives the same data after a seasonal adjustment to remove the quasi-regular seasonal cycle. The adjustment involves subtracting from the data a 4-harmonic fit with a linear

[5] http://scrippsco2.ucsd.edu/data/atmospheric_co2/mlo

Table 7.1 *The first five rows of monthly averaged CO_2 concentrations (ppm) measured at Mauna Loa Observatory*

Yr	Mn	Date Excel	Date	CO_2 [ppm]	Seasonally adjusted [ppm]	Fit [ppm]	Seasonally adjusted fit [ppm]	CO_2 filled [ppm]	Seasonally adjusted filled [ppm]
1958	1	21200	1958.0411	–99.99	–99.99	–99.99	–99.99	–99.99	–99.99
1958	2	21231	1958.126	–99.99	–99.99	–99.99	–99.99	–99.99	–99.99
1958	3	21259	1958.2027	315.69	314.43	316.19	314.9	315.69	314.43
1958	4	21290	1958.2877	317.46	315.15	317.3	314.98	317.46	315.15
1958	5	21320	1958.3699	317.51	314.73	317.84	315.06	317.51	314.73

Data from http://scrippsco2.ucsd.edu/data/atmospheric_co2/mlo

> gain factor. Column 7 is a smoothed version of the data generated from a stiff cubic spline function plus 4-harmonic functions with linear gain. Column 8 is the same smoothed version with the seasonal cycle removed. Column 9 is identical to Column 5 except that the missing values from Column 5 have been filled with values from Column 7. Column 10 is identical to Column 6 except missing values have been filled with values from Column 8. Missing values are denoted by –99.99.

The metadata tells us that Columns 1–4 are redundant: these columns provide the age scale of the data in a number of different formats. Column 1 is year and Column 2 is month, where January is 1, February is 2 and so on. Column 3 is specifically for Excel users and provides the Excel code that refers back to the date and month from Columns 1 and 2. Column 4 is decimal year, with the days of the year being converted to a decimal. Depending on the software you are using to plot the data, you might choose to use any of these columns for your timescale.

The metadata also tells us that Column 5, CO_2 in ppm (parts per million) is the most commonly used CO_2 data set. The other columns are the same CO_2 data but they have been processed in different ways. My intention right now is to demonstrate some of the issues that arise with raw data, so I am going to choose the closest to raw, unprocessed data that I can get. So, I will choose Column 5 as my working data set. When data is available in different formats, choose the format that is the most useful for your purpose.

Finally, the metadata says that missing data is denoted by –99.99. This is very important information. It means that whenever there is a –99.99 in the spreadsheet, there is, in fact, no data for that cell. When I look at the spreadsheet, I can see –99.99 in the first two rows, indicating that the data collection didn't start until March of 1958. If you skimmed the full spreadsheet, you would see other instances of missing data. It is important to pay attention to this missing data as you proceed with your analysis.

7.4 STEP TWO: PLOTTING THE DATA

In order to get a feel for data, you need to plot it. It is very difficult to get an intuitive sense for what is going on – the patterns, cycles, and events in the data – just by looking at a long list of numbers.

I will plot the data in Excel. Rather than choosing Column 3 – the excel code for dates – I am going to choose Column 4, decimal year. This is because I get no intuitive understanding of the *x*-axis when it is plotted as numbers rather than years. I want to be able to see the timescale associated with the cycles in the data. If there is a unique event, I want to know when exactly it happened in time. I will plot Columns 4 and 5. If you are using a different plotting programme, pick the columns that are most useful for you to clearly see the dates on the *x*-axis.

Now I have to decide what type of plot to generate. (To remind yourself about different approaches to plotting data, revisit Chapter 2.)

What is the best choice for this data?

I am interested in looking at the change in CO_2 over time; therefore I want a graph that will have time on the *x*-axis and CO_2 on the *y*-axis. A scatter plot will allow me to plot this timeseries effectively. If you decide to connect the datapoints with a line (which would be fine in this case because the samples are sequential in time), choose the most simple, straight-line connection. The curved connection lines in Excel generate smoothed out transitions. If you do not know anything about the transition between one data point and the next, do not assume that the transition was curved. The linear connection more directly represents the lack of knowledge about the transition between datapoints.

When I plot the data from Columns 4 and 5 as a scatter plot (Figure 7.2), choosing both markers for the datapoints and a line connecting the markers but without doing anything else (i.e., using the default settings), this is what I get:

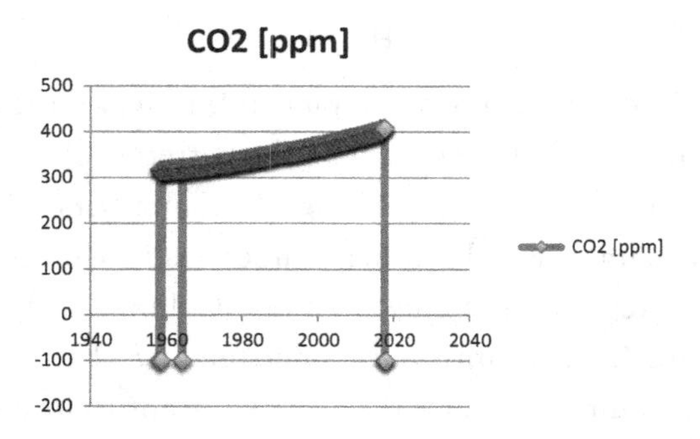

FIGURE 7.2 Mauna Loa Atmospheric CO_2 (ppm) plotted using default settings only.
Data from http://scrippsco2.ucsd.edu/data/atmospheric_co2/mlo.

This is not a helpful figure.

Recall from the metadata that –99.99 is used to indicate no data. If you plot the data with the –99.99 placeholders included in the spreadsheet, you will be unable to see the real data properly. You will be plotting an artefact, or introduced element, of the data analysis. To plot the data properly, you will need to remove the –99.99s from the spreadsheet.

For management purposes, always keep the raw, unprocessed data as a separate spreadsheet. Make a copy to be your "working data set." This allows you to always go back and check that what you are seeing makes sense and is, in fact, an element of the original data and not an artefact.

Depending on the software you are using, you can rename the –99.99s as Not a Number (NAN), or you can "find" the –99.99s and "replace" them with blank cells. You do not want to replace the cells with another number, like 0, that will introduce problems when calculating averages or correlations.

When plotting your graph, do not rely on the default settings of your software programme: they will hardly ever suit your needs. You will probably have to change or define the scale to fit the range of data that you need. Excel tends to format all plots to include 0 on the *y*-axis. In this example, with atmospheric CO_2 ranging from 313 ppm to over 400 ppm, including 0 on the *y*-axis decreases our ability to see the interesting variability in the data. You want both the *x*- and *y*-axes to fit the data set so you can see the details of what is happening quite easily. Adjust the axes so the "white space" around your data set is minimal.

I am aiming to generate a plot that captures both the increasing trend in the CO_2 data over time and the annual and sub-decadal cycles of variability that exist. In Figure 7.3a, the CO_2 data is plotted only as points – with no connecting line. I know that this image will be printed in greyscale, not in colour. To plot the greyscale value that I want, I need to change the colour of the markers. I also want the image to be simple and clear. To generate this plot, I changed

- both the *x*- and *y*-axes scales,
- the colour and the size of the markers,
- the colour and width of the line, and
- the size of the label fonts.

I removed the shadow behind the marks, grid lines, the legend, and the title: all were unnecessary for this graph. Finally, I added labels to the *x*- and *y*-axes.

Figure 7.3a gives you an overall indication of the long-term trend in the data. The data points fall in a thick band that is continually increasing over time. Depending on the size of your data point markers (in Figure 7.3a they are 6 pt) you will either see only the overall increase or, if your points are small enough, you might see annual cycles as well.

In Figure 7.3b I have replotted the same data using only a line. The line graph allows the annual cycle to be more evident. Comparing Figure 7.3a and b, it is clear that the way you choose to plot the graph will determine the information that readers will receive. Do not expect that your programme or software can determine the best presentation of your data. You need to spend some time working with the data to decide on the best presentation for your particular purposes.

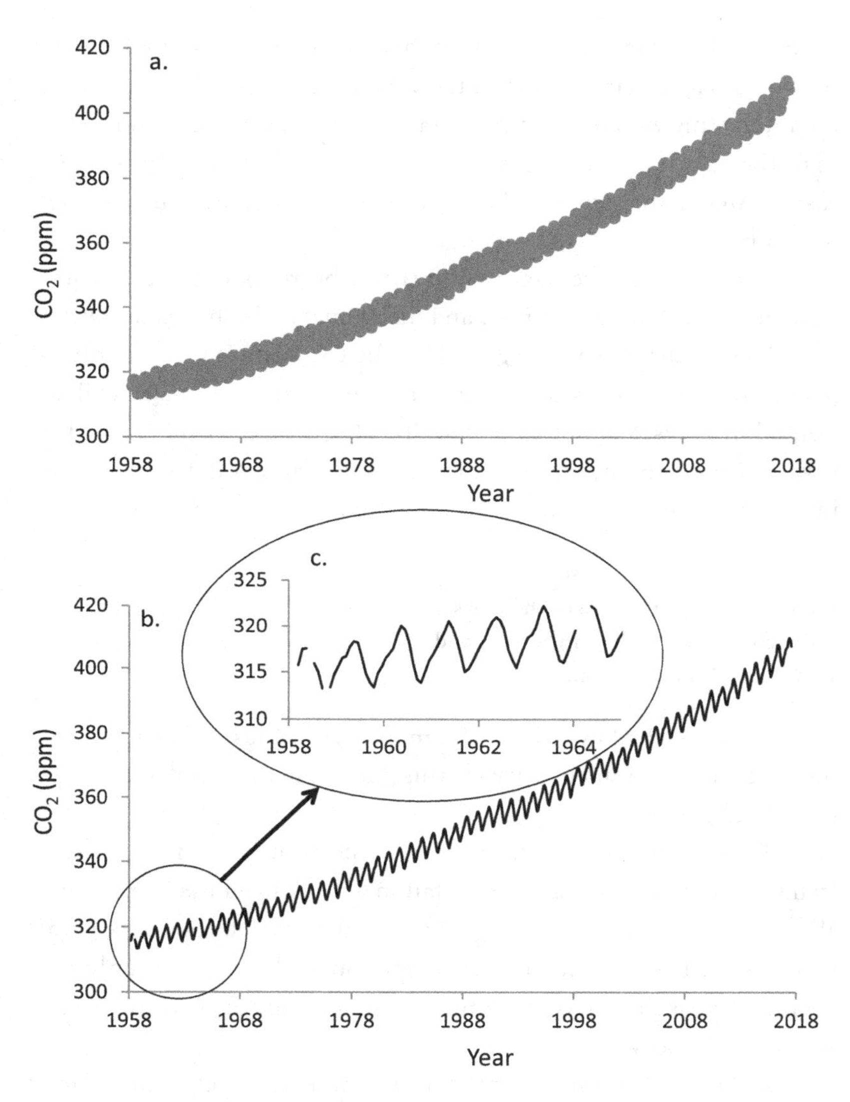

FIGURE 7.3 The concentration of atmospheric CO_2 (ppm) measured at Mauna Loa Observatory, plotted two different ways. (a) Average monthly atmospheric CO_2 (ppm) data plotted using markers for every datapoint. (b) Average monthly atmospheric CO_2 (ppm) data plotted without markers, using only a line to connect datapoints. (c) Average monthly atmospheric CO_2 (ppm) data from 1958 to 1965 only.

Data from http://scrippsco2.ucsd.edu/data/atmospheric_co2/mlo.

7.5 STEP THREE: FILLING DATA GAPS

Remember that there were some missing data in the spreadsheet (denoted by −99.99). I will need to address these data gaps before I can start working with the data.

There are a number of different ways to fill data gaps. To decide which way is best for the situation, I need to have a look at the data to see where the gaps fall. If I cross-reference between the raw data and Figure 7.3c, I can see that Gap 1 is made up of two missing data points at the very beginning of the year, so the CO_2 data doesn't actually start until March. Gap 2 is only one data point and it falls just after the peak value in 1958. Gap 3 is one data point and it falls at the lowest part of the cycle in 1958. Gap 4 is three data points, falling just before the peak value in 1964. Each gap has a slightly different challenge. I will have to consider each gap separately.

Gap 1: This is not really a gap. Though the spreadsheet starts in January, the data collection did not start until March. I am not going to try to generate data before the data collection started. The process of adding data to the beginning or end of a data set is called extrapolating. You need to have pretty solid reasons to extrapolate data. In this case the only reason is aesthetic – starting in January rather than March – so I will not attempt to fill this "non" gap.

Gap 2: Adding data within an existing sequence is called interpolating. I can interpolate data for Gap 2 because there is a fairly consistent pattern evident in the data. Looking at the full record, I can see that there is a gradual rise and a gradual decrease in the CO_2 values every year. This missing value for Gap 2 is not at the maximum or minimum position of the cycle. Therefore, it is a fair assumption to say that the missing data point will fall between the two bracketing points. I will average these two points to generate a reasonable value to fill Gap 2.

Gap 3: Gap 3 is a more difficult gap to fill because it occurs at a minimum value in the sequence. I cannot average the two bracketing values because doing that will underestimate the minimum value and not accurately reflect the cycle amplitude. If I underestimate the cycle amplitude, I cannot calculate an accurate annual average.

Given the position of Gap 3, I might reconsider using the first year of data. I might decide that 1958 has too many gaps to accurately generate an annual average for that year. I would then truncate the data set and start my analysis in 1959, the first complete year of data collection. In light of my question, and given that we have 58 years of data, this would be a reasonable approach to take.

However, if you really needed that first year of data, how could you reasonably fill Gap 3?

Before deciding on my approach, I would like to get a feel for how the minimum values of the data set are behaving. To do this, I will isolate the lowest values of every year (October values) and then plot them alone. Figure 7.4a gives me a much better understanding of how the October values (the minimum values) are changing over time. I can see that from 1959 to 2017, the increase in the October values is not linear. But I can also see that I could approximate the change in the first decade of the data set using a straight line.

To extrapolate a datapoint for October 1958, I could fit a straight line through the existing October values for 1959–1968 and then extend that line to find an appropriate value for October 1958. I do not, in fact, know how the atmospheric CO_2 concentrations for October of 1958 behaved. It could have been an anomalous year. But I can assume that it was not an anomalous year and then go ahead and fill Gap 3 with an extrapolated value.

How do I go about drawing a line through the data points?

I would like to find the line that best "fits" my data. The line with best fit is the line that has the minimum distance possible between the datapoints and the line. For a given x value, the difference between the observed y value (your datapoint) and the predicted y' value, based on the line of best fit, is called a residual (Figure 7.5). Of all possible lines, the line of best fit is the line that minimizes the sum of the squared residuals. (Squaring the residuals makes all of the values positive.)

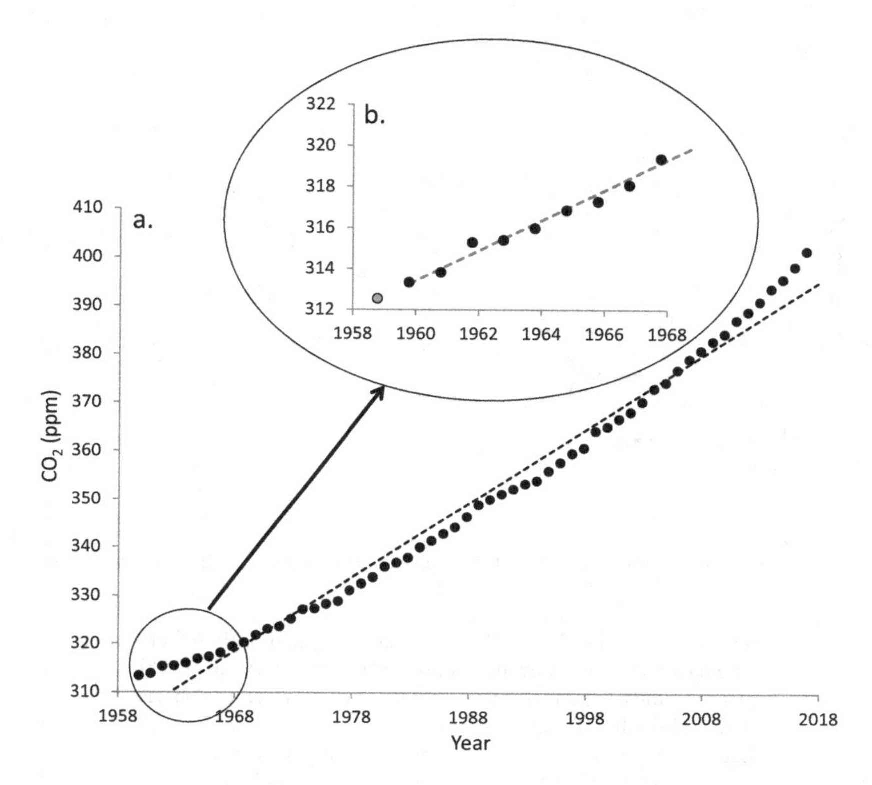

FIGURE 7.4 Atmospheric CO_2 (ppm) data for the month of October only. (a) October CO_2 concentrations (ppm) from 1958 to 2017. The dashed line is a linear fit of the data. (b) October CO_2 concentrations (ppm) from 1958 to 1968. The dashed line is a linear fit of the data. The gray marker is an extrapolated data point based on the line of best fit.
Data from http://scrippsco2.ucsd.edu/data/atmospheric_co2/mlo.

The least squares, or linear regression, method is used to generate a straight line with a slope (m) and intercept (b) that minimizes the sum of the squared residuals:

$$y' = mx + b$$

where

$$m = \frac{\sum(x_i - x_{avg})(y_i - y_{avg})}{\sum(x_i - x_{avg})^2}$$

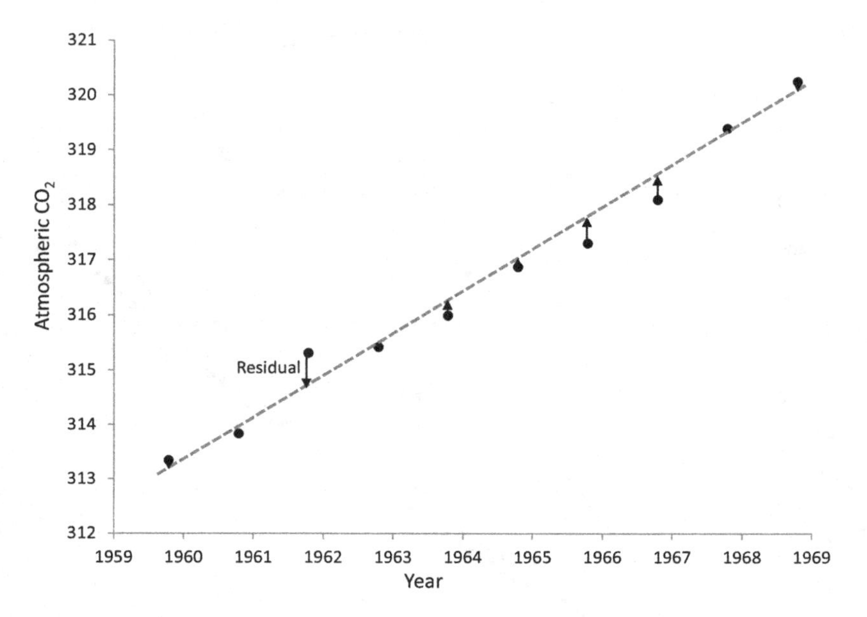

FIGURE 7.5 Atmospheric CO_2 (ppm) data for the month of October only. The dashed line is an eyeballed line of fit through the data. The black arrows indicate the residuals: the difference between the observed CO_2 data and the line.

Data from http://scrippsco2.ucsd.edu/data/atmospheric_co2/mlo.

and

$$b = y_{avg} - mx_{avg}$$

Using a spreadsheet can be helpful for complicated calculations. Use a separate column or cell for each step in the process (Figure 7.6). This lets you follow your work and find problems as they arise. Of course, there is a linear regression function in all statistics programmes. The challenge is to ensure that you can actually do the calculation yourself before relying on the black box options provided by your software.

Based on the least squares calculations, the line that best fits the CO_2 data from 1959 to 1968 is

$$y' = 0.74x - 1130.79$$

	A	B	C	D	E	F	G	H	I
1	Avg X =	=AVERAGE(C2:C11)	X = Date	Y = CO2 (ppm)	X-AvgX	(X-AvgX)^2	Y-avgY	(X-AvgX)*(Y-avgY)	Regression
2	Avg X =	1964.289	1959.789	313.34	=C2-$B$2	=E2^2	=D2-$B$5	=E2*G2	=$B$8*C2+$B$11
3			1960.790	313.83	-3.500	12.247	-2.75	9.62	314.00
4	Avg Y =	=AVERAGE(D2:D11)	1961.789	315.31	-2.500	6.251	-1.27	3.18	314.74
5	Avg Y =	316.58	1962.789	315.42	-1.500	2.251	-1.16	1.74	315.47
6			1963.789	315.99	-0.500	0.250	-0.59	0.30	316.21
7	m = slope =	=H15/F15	1964.790	316.87	0.500	0.250	0.29	0.15	316.95
8	m = slope =	0.7368	1965.789	317.30	1.500	2.249	0.72	1.08	317.69
9			1966.789	318.10	2.500	6.249	1.52	3.80	318.42
10	b = intercept =	=B5-(B8*B2)	1967.789	319.39	3.500	12.249	2.81	9.83	319.16
11	b = intercept =	-1130.7892	1968.790	320.25	4.500	20.254	3.67	16.52	319.90
					Sum =	=SUM(F2:F11)	Sum=	=SUM(H2:H11)	
					Sum =	82.502	Sum=	60.79	

FIGURE 7.6 Calculating a line of best fit for average October CO_2 concentrations from 1959 to 1968 using a spreadsheet.
Data from http://scrippsco2.ucsd.edu/data/atmospheric_co2/mlo.

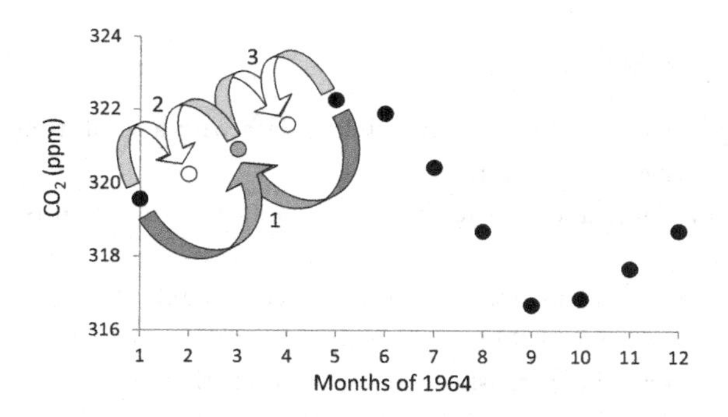

FIGURE 7.7 Cartoon demonstrating the order of steps taken to interpolate three CO_2 data points from 1964: (1) Averaging January and May data to generate a March value. (2) Averaging January and March data to generate a February value. (3) Averaging March and May data to generate an April value.
Data from http://scrippsco2.ucsd.edu/data/atmospheric_co2/mlo.

Substituting 1958 for x in this equation, $y' = 312.53$. Gap 3 can now be filled.

Gap 4: Gap 4 has more missing values than our previous examples, three, in fact. They are associated with the February, March, and April values for 1964. I can take a first stab at filling this gap by averaging the bracketing values (January and May) to calculate a value for March (Figure 7.7). Now that I have a March value, I can use it to generate a value for February by averaging the actual data value for January with my calculated value for March. Finally, I can also use the calculated

March value, along with the actual data for May, to calculate a value for April.

Filling data gaps is not trivial. The four examples above demonstrate some different approaches to dealing with data gaps. I would suggest taking an approach that matches your needs and your skill set.

Remember to keep track of how you manage missing data. At some point, when writing about your process, you will have to report your data management methods. I have now filled all of the critical gaps in the data set and there are no longer any placeholders present. I am now ready to proceed with my data analysis.

7.6 TAKE-HOME MESSAGES

- When using previously published data, choose data with a resolution that allows you to answer your question.
- When data is available in different formats, choose the format that is useful for your purpose.
- Read and save the metadata along with the raw data as archive and resource.
- Do all data analysis on a working spreadsheet, not on the raw data archive.
- Take time to set up your spreadsheets using columns that are useful to you.
- Remove data placeholders and fill gaps based on your needs and skill set.
- When plotting your data, change the default settings (marker size, colour, line width, scale, etc.) to meet your needs.

REFERENCES

Ahn, J., Brook, E. J., Mitchell, L. et al. (2012). Atmospheric CO2 over the last 1000 years: A high-resolution record from the West Antarctic Ice Sheet (WAIS) Divide ice core. *Global Biogeochemical Cycles* 26: GB2027. https://doi.org/10.1029/2011GB004247

Barnola, J. M., D. Raynola, Y. S. Korotkevich, and C. Lorius (1987). Vostok ice core provides 160,000-year record of atmospheric CO_2. *Nature*, 329: 408–414.

Bereiter, B., S. Eggleston, J. Schmitt et al. (2015). Revision of the EPICA Dome C CO_2 record from 800 to 600 kyr before present. *Geophys. Res. Lett.*, 42: 542–549. DOI: 10.1002/2014GL061957.

Etheridge, D. M., L. P. Steele, R. L. Langenfelds et al. (1996). Natural and anthropogenic changes in atmospheric CO_2 over the last 1000 years from air in Antarctic ice and firn. *J. Geophys. Res.*, 101: 4115–4128.

Keeling, C. D., S. C. Piper, R. B. Bacastow et al. (2001). Exchanges of atmospheric CO_2 and $13CO_2$ with the terrestrial biosphere and oceans from 1978 to 2000. I. Global aspects, SIO Reference Series, No. 01-06, Scripps Institution of Oceanography, San Diego, 88pp.

Luthi, D., M. Le Floch, B. Bereiter et al. (2008). High-resolution carbon dioxide concentration record 650,000–800,000 years before present. *Nature*, 45: 379–382. DOI: 10.1038/nature06949.

Siegenthaler, U., T. F. Stocker, E. Monnin et al. (2005). Stable carbon cycle–climate relationship during the Late Pleistocene. *Science*, 310: 1313–1317.

Chapter 8

Preparation

1. Before reading Chapter 8, consider the following questions:
   - How can one signal in a data set be isolated?
   - What is the difference between an average and a running average?
   - What is an anomaly data set?
2. Imagine how the number of students on your campus changes over a day, a year, and a decade. Draw a graph for each timescale.

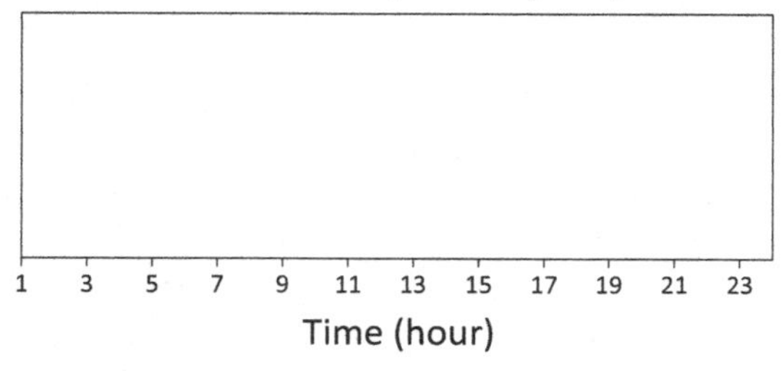

Students on X campus: Daily Snapshot

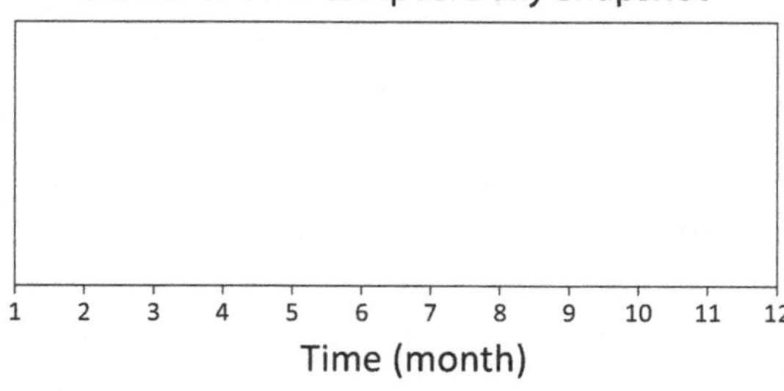

Students on X campus: Decadal Snapshot

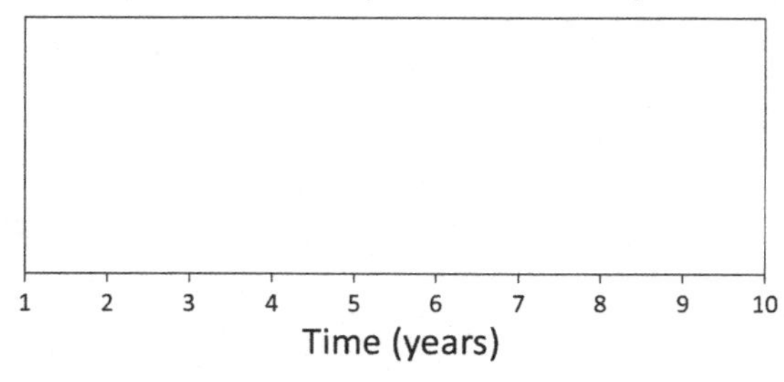

3. Draw a new graph, using only one line, that captures (combines) the three different timescales of change that you have sketched out in question 2.

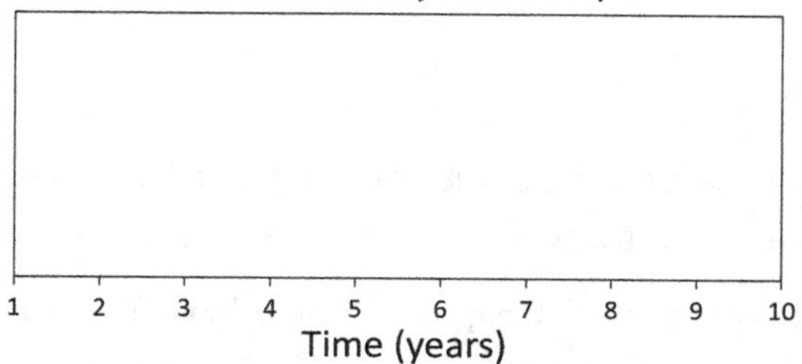

8 Isolating Individual Signals from a Composite Data Set

8.1 WHAT SIGNALS ARE PRESENT IN THE MAUNA LOA ATMOSPHERIC CO_2 DATA SET?

After completing the steps in Chapter 7, I now have a well-prepared data set of monthly averaged atmospheric CO_2. The placeholders have been removed and the gaps have been filled. I can now proceed to answer my original question:

How have concentrations of atmospheric CO_2 changed since 1958 as a result of natural variability and human CO_2 emissions, respectively?

Now that the data set is complete, without gaps, a scatterplot line graph clearly shows the details of the data set (Figure 8.1a).

What part of this graph indicates the human input to atmospheric CO_2 and what part represents the natural variability?

There is a lot of information published about atmospheric CO_2 investigating the first-order processes that influence atmospheric CO_2 concentrations. To help you make decisions as you analyze the data, gather more background information or refer to peer-reviewed papers that you might already have collected. The scientific consensus is that the consistent increase in atmospheric CO_2 concentrations over time reflects the anthropogenic, or human generated, signal caused from burning fossil fuels. The annual cycle is driven by plant photosynthesis, which consumes atmospheric CO_2, and plant respiration, which releases CO_2 to the atmosphere. The net effect of these processes removes CO_2 from the atmosphere during the growing season and releases CO_2 back to the atmosphere during the dormant season.

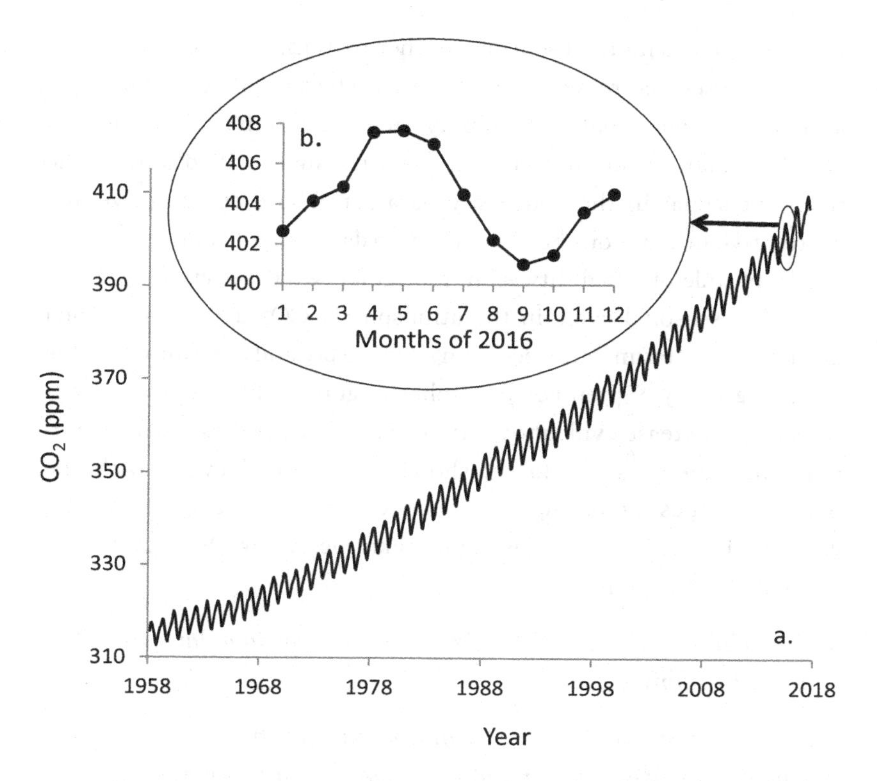

FIGURE 8.1 Monthly average atmospheric CO_2 concentrations (ppm) measured at Mauna Loa Observatory, Hawaii. (a) Monthly average atmospheric CO_2 concentrations (ppm) from 1958 to 2018. (b) One annual cycle (2016) of monthly averaged CO_2 measurements.
Data from Keeling et al. (2001). http://scrippsco2.ucsd.edu/data/atmospheric_co2/mlo.

Though both the anthropogenic and biological processes influence the concentration of atmospheric CO_2, they are different processes. The human activities that generate CO_2 – mainly the burning of fossil fuels – are not related to the biological process that generates CO_2 – respiration.

The Mauna Loa atmospheric CO_2 data, therefore, has (at least) two signals superimposed on top of each other: an annual cycle and a long-term trend. Theoretically, if you add data representing one pattern to data representing another pattern, the result is a composite data set that reflects both patterns. The two additional positive (or two negative) signals will amplify each other in the composite data set. The addition of signals with opposite signs will diminish the resultant signal in the composite data set. These constructive and destructive interactions between the two data sets add or detract from the amplitude of variability seen in original two data sets.

The seasonal cycle in the atmospheric CO_2 data from Mauna Loa has a maximum value occurring every May and a minimum value occurring every September / October (Figure 8.1b). There is also a consistent increase evident over time. The result of these two concurrent phenomena is a data set showing an annual cycle, with the monthly values increasing from year to year: the average May CO_2 value for 1959 (318.29 ppm) is higher than the average May CO_2 value in 1958 (317.51 ppm).

So how can signals be isolated from one another in a data set comprising many signals?

How can the seasonal cycle of atmospheric CO_2 be studied independently from the consistent increase in atmospheric CO_2 over time?

8.2 CALCULATING AVERAGES

An environmental data set with many superimposed, but non-interacting, signals can theoretically be decomposed into individual signals by averaging to remove a high-frequency signal and then subtracting the average from the raw data. How simple or complicated this is depends on how distinct the signals are from one another. In the example of the atmospheric CO_2 data, the two different signals are distinct. Therefore, I can first average the raw data and then subtract the average (in this case representing the anthropogenic signal) from the raw (composite) data set to isolate the annual cycle.

To generate a set of numbers that represents the increase over time, I am going to average the data. *But how?*

If I generate an annual average, by averaging the data for all 12 months of one year and then all 12 months of the next year and so on, I will generate a new data set with one data point representing each year (Figure 8.2). Taking this approach, I will reduce a data set with monthly resolution (Figure 8.2a) to one that

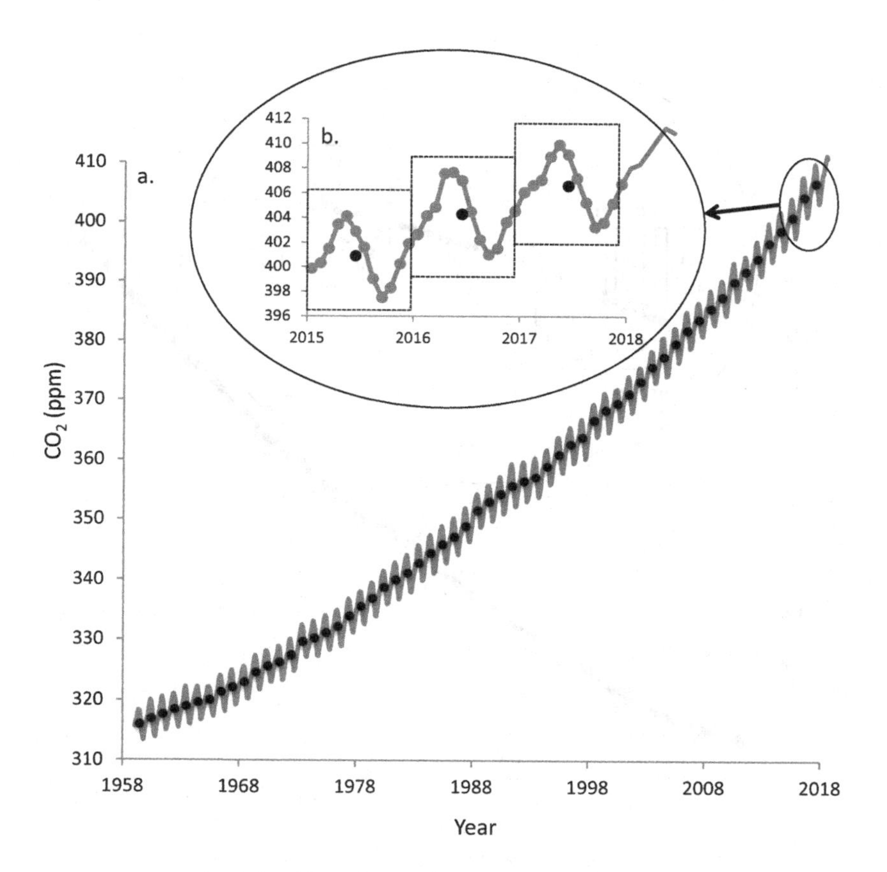

FIGURE 8.2 Monthly and annual average atmospheric CO_2 concentrations (ppm) measured at Mauna Loa Observatory, Hawaii. (a) CO_2 data from 1958 to 2018 is presented. Monthly averages are plotted in grey. Annual averages are plotted as the black circles. (b) CO_2 data from 2015 to 2018 is presented. Monthly averages are plotted in grey. Annual averages are plotted as the black circles. The squares isolate the data points from each year used to generate the annual average.

Data from http://scrippsco2.ucsd.edu/data/atmospheric_co2/mlo.

has annual resolution (Figure 8.2b) and I will lose a fair bit of information.

Data is precious and I don't want to lose resolution unnecessarily. So, instead of an annual average for each year, I will generate an average of 12 months with a staggered start month (Figure 8.3a).

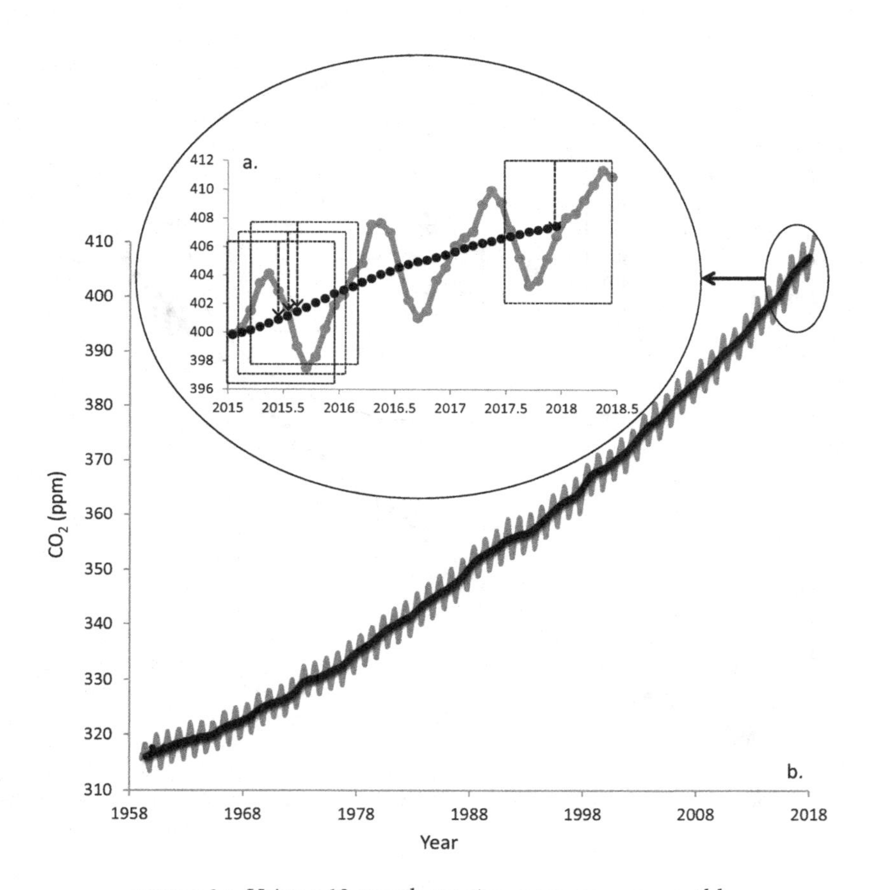

FIGURE 8.3 Using a 12-month running mean average monthly atmospheric CO_2 concentrations (ppm). (a) CO_2 data from 2015 to 2018 is presented. Monthly averages are plotted in grey. The 12-month running mean is plotted as the black circles. The squares isolate the 12 data points used to generate the running average. (b) CO_2 data from 1959 to 2018 is presented. Monthly averages are plotted in grey. The 12-month running mean is in black.

Data from http://scrippsco2.ucsd.edu/data/atmospheric_co2/mlo.

I can start in January and average data for 12 months – January to December. Then I can average data from February to January, and then March to February and so on. This 12-month "running average" maintains the monthly resolution of the data but smooths the data set to remove the variability on timescales of one year and less.

The amount of time over which data is averaged is called the "averaging window." To isolate the long-term variability present in the atmospheric CO_2 data, the annual variability needs to be removed. Averaging over a 12-month window will remove variability that occurs on a timescale less than 12 months, but it will not remove variability with a time scale more than 12 months. If you were using a different data set, for example one that had a strong El Niño / Southern Oscillation (ENSO) signal (cycle of three to five years), then you would have to average over a window of five years or more to remove the ENSO signal from the data. The averaging window should match the timescale of the variability that needs to be removed.

When calculating an average, what x-axis value should I pair with the new y-axis value?

Should I align the first value of the new smoothed data set in the cell associated with the first data point that I averaged or with the last data point that I averaged or somewhere in between?

This is just a practical decision, but one that impacts the way data is viewed in a graph. If the new data set starts in the cell at the beginning of the averaged range, there will be no data points for the most recent timeframe of the graph. If the new data set starts in the last cell of the averaged window, there will be no data points associated with the beginning of the range. Of course, the data is there – it has been included in the averaging – but visuals make an impact. By convention, the first datapoint of a running average is usually placed half way through the averaged values. As I used a 12-month running mean (averaging an even number of values), I placed the first new averaged value in month 6 (June) of the first year.

Figure 8.3b shows the new running average on the same graph as the original data set. The running average generates a line in the centre of the original data set tracking the average increase in atmospheric CO_2 over time.

8.3 CALCULATING RESIDUALS

Are we there yet? Almost.

The annual running average of the Mauna Loa atmospheric CO_2 data reflects the human contribution to atmospheric CO_2 since 1958. But I have not yet isolated the annual cycle that reflects the natural variability driven by photosynthesis and respiration.

To isolate the annual cycle, we need to remove, through subtraction, the trend from the full data set. In this action, we are generating an anomaly data set of residuals (or non-dimensional data). The anomaly CO_2 data set now varies around a mean of 0 and shows deviations above and below the mean (Figure 8.4). The units presented now are not the actual values of the concentration of CO_2 in the

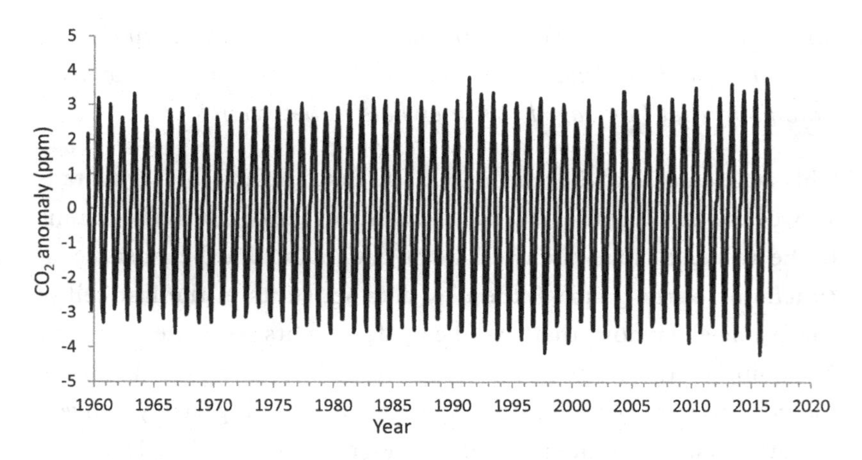

FIGURE 8.4 CO_2 anomalies from 1960 to 2018. This data set was generated by removing a 12-month running mean from the Mauna Loa Monthly averaged CO_2 data set.

Data from http://scrippsco2.ucsd.edu/data/atmospheric_co2/mlo.

atmosphere every month, but are instead representative of monthly deviations from the mean.

Though the anomaly plot (Figure 8.4) looks quite a bit different from the original (Figure 8.1a), it is important to remember that all we have done is isolate a signal that was always present in the original data set. We have not changed the information; we have just focused on one component of a complicated natural signal.

The plot of residuals (Figure 8.4) highlights an annual cycle with maximum values in May and minimum values in September / October. Check these against the original data set to make sure that the peaks and troughs are occurring at the same time in the new data set as they were in the original data set. If you see that your residual plot has features that were not present in the original data set, double check to ensure that your calculations were correct. If, for example, you subtracted the full data set from the calculated running average, your maximum and minimum values would be reversed: the maximum value would occur in October and the minimum value would occur in May. This would be an error. Your data analysis should not generate artefacts, or manifest features, that were not present in the original data.

8.4 DESCRIBING THE RESULTS

My research question asked:

How have concentrations of atmospheric CO_2 changed since 1958 as a result of natural variability and human CO_2 emissions, respectively?

The plot of CO_2 residuals, generated from removing the 12-month running average from the Mauna Loa record of atmospheric CO_2, effectively isolated the non-anthropogenic contribution to atmospheric CO_2. When describing results, it is important to report both the total amount of change and the range over which the change occurs. Clearly describe all scales of change that are evident. Your verbal description of data should be detailed and specific enough that a reader could sketch the graph that you have produced without looking at it.

So, I can now answer my research question:

Monthly averaged atmospheric CO_2 concentrations, collected from Mauna Loa, Hawaii, were analyzed to isolate two constituent components: (1) the natural cycle of photosynthesis and respiration; (2) the long-term increase associated with human activity.

At this location, the burning of fossil fuels has resulted in a progressive increase in the annual average concentration of atmospheric CO_2, from 316 ppm in 1959 to 406.5 ppm in 2017. Human activity between 1959 and 2017 resulted in an average increase in atmospheric CO_2 of 90.6 ppm.

Since 1959, the average annual variability of atmospheric CO_2, driven by plant photosynthesis and respiration in Hawaii, has been 6.4 ppm with the maximum values occurring in May and the minimum values occurring in September / October.

8.5 MORE INFORMATION GENERATES MORE QUESTIONS

Creating Figure 8.4 revealed features that were not easily identifiable in the original data set. I reported average values in my summary earlier. But looking closely at Figure 8.4, it appears that the seasonal cycle is not exactly the same every year. Notice that some years have larger amplitudes than other years. The maximum values of CO_2 in the atmosphere over Mauna Loa appear to be increasing over time, while the minimum values appear to be decreasing over time. Figure 8.4 could be used to prompt more questions.

How has the seasonal cycle changed since 1959? Are the annual maximum and minimum vales in atmospheric CO_2 changing over time?

To investigate the variability of the annual cycle, I could

- compare the monthly averages over the timeframe of the full data set;

- investigate whether there has been a change in the amplitude of the annual cycle over time from 1959 to 2017; and
- determine whether there has been a unidirectional trend in maximum and minimum CO_2 values observable since 1959.

Our initial question has now generated at least three more lines of investigation to dig deeper into the same phenomena.

8.6 TAKE-HOME MESSAGES

- It is possible to identify and isolate individual, non-interactive signals from a composite data set.
- Running averages can be used to isolate low frequency variability by smoothing out higher-frequency variability without losing data resolution.
- The "averaging window" must be chosen to match the frequency of variability that is to be removed.
- Isolating a signal is a way to highlight specific information that is already present in the data.

REFERENCE

Keeling, C. D., S. C. Piper, R. B. Bacastow et al. (2001). Exchanges of atmospheric CO_2 and $13CO_2$ with the terrestrial biosphere and oceans from 1978 to 2000. I. Global aspects, SIO Reference Series, No. 01-06, Scripps Institution of Oceanography, San Diego, CA, 88pp.

Chapter 9

Preparation

1. Before reading Chapter 9, consider the following questions:
   - When looking at environmental data, how can you tell what is a signal and what is noise?
   - When more than one pattern of variability is present in the data set, how can you decide which one is dominant?
   - If there are multiple signals of variability present in the data, can we always isolate them?
2. You have been asked to isolate the annual pattern of air pollution in a particular location from a record of ozone concentrations that has an hourly resolution collected over 20 years.
   a. Do you think you could do it?
   b. How would you try to go about doing it?
   c. What might complicate your ability to isolate the annual variability in the record?

9 Differentiating Signals from Noise

9.1 SIGNAL DOMINANCE

In Chapter 8, the Mauna Loa CO_2 data from 1958 to 2017 was decomposed to isolate the annual signal from the long-term trend. That exercise demonstrated that two non-interacting cycles, existing in the very same data set, can be considered separately. Once isolated, the amplitude of each cycle can be compared to identify the relative dominance of the different signals. Taking this to heart, a conceptual understanding of time series analysis can be developed.

To investigate signal dominance, or relative signal strength, I will work through two examples, each comprising three different signals: random noise, a medium-frequency cycle, and a linear increase over time. Random noise means that there is no relationship between one datapoint and the next datapoint; knowing one data point does not allow the next datapoint to be predicted. In both the examples, I generated the random noise data sets using the =norminv(probability,mean,std) function in Excel with "rand" as the probability. In Example 9.1, the three different signals will have three different amplitudes. In Example 9.2, the amplitudes of the random noise and the medium frequency variability will be similar.

EXAMPLE 9.1 **Consider three different types of variability**

A. low amplitude noise
B. cyclical medium frequency, medium amplitude variability
C. linear trend

If signal A has an amplitude of <1, signal B has an amplitude of ~2, and signal C increases 1.5 units over 10 years, what would the resultant combined data set look like, where all three signals are present at the same time?

To answer this question, I have fabricated three different data sets that match the three different cycles of variability outlined above (A, B, C). When plotted separately (Figure 9.1A–C), the amplitude of change and the timescale associated with each cycle are clear. When these three different signals are added up – literally summed – the result is a graph with one line that captures all of the cycles in the data set superimposed on top of one another (Figure 9.1D). If you look for them, you can still see the three different signals. But the composite data set reflects the relative strength of the three signals.

Which signal dominates the composite record?

The dominant signal is the signal that has the highest amplitude over the timeframe in question. The timeframe determines the "window" through which you are looking at the data. The timeframe will influence your answer to the question of signal dominance. Remember, the timescale doesn't determine dominance: the signal with the largest timescale is not automatically the dominant signal. It is the relative amplitude that determines signal dominance.

In Figure 9.1, the fabricated data set is plotted over 10 years. In Figure 9.2, the same data is plotted over 100 years. Compare the amplitudes and timescales for the three different signals over the two different snapshots in time:

- Over a 10-year timeframe (Figure 9.3a), the largest amplitude is associated with signal B. The medium-frequency signal cycles with an amplitude of ~2. The random noise (A) has an amplitude of ~0.6 and the long-term trend (C) only increases ~1.5 units over 10 years. In relative terms, over 10 years, the medium-frequency signal (B) has the highest amplitude, and is therefore dominant.
- Over a 100-year timeframe (Figure 9.3b), the amplitude of the linear trend (C) is ~15, while the amplitude of signal (B) still remains ~2 and

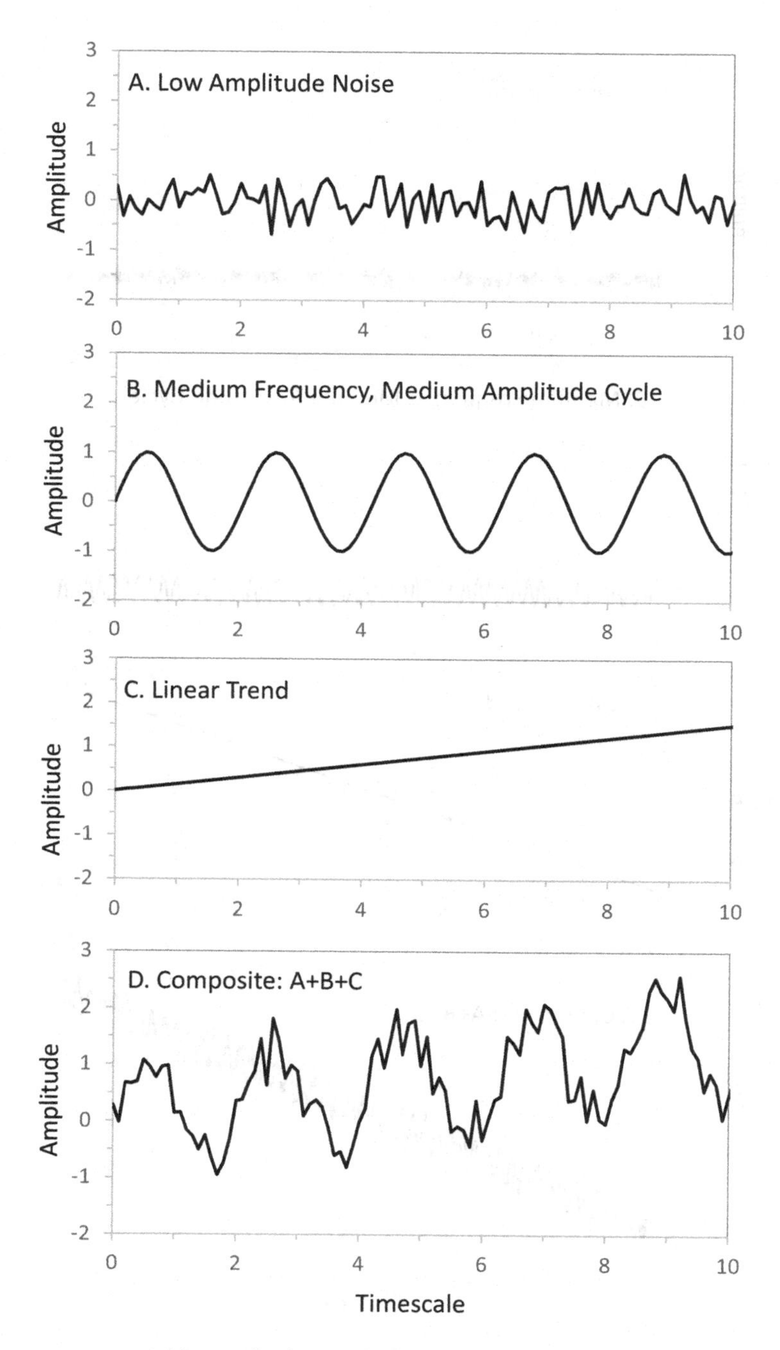

FIGURE 9.1 Adding multiple signals together to generate a composite data set: 10-year timeframe. The three signals, A, B, and C, were added together to generate D.

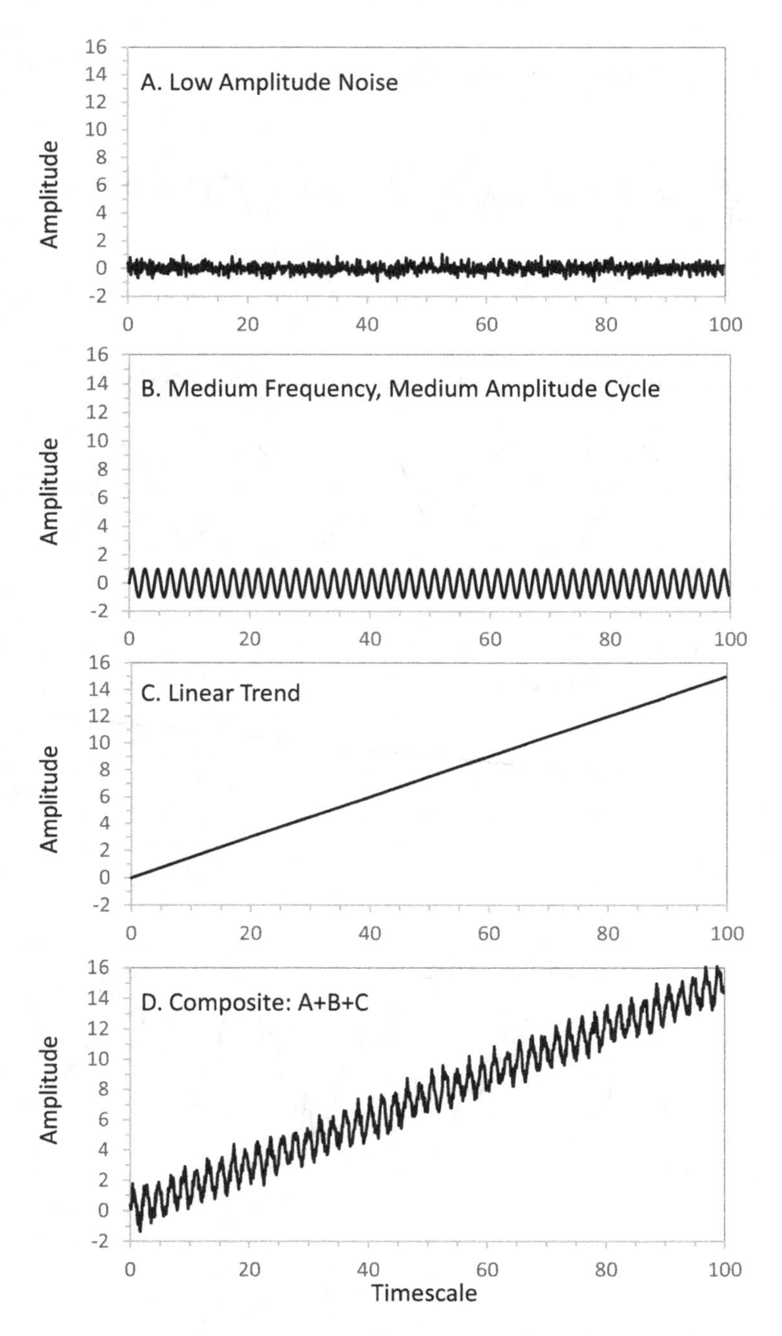

FIGURE 9.2 Adding multiple signals together to generate a composite data set: 100-year timeframe. The same three signals from Figure 9.1 are presented over a longer period of time.

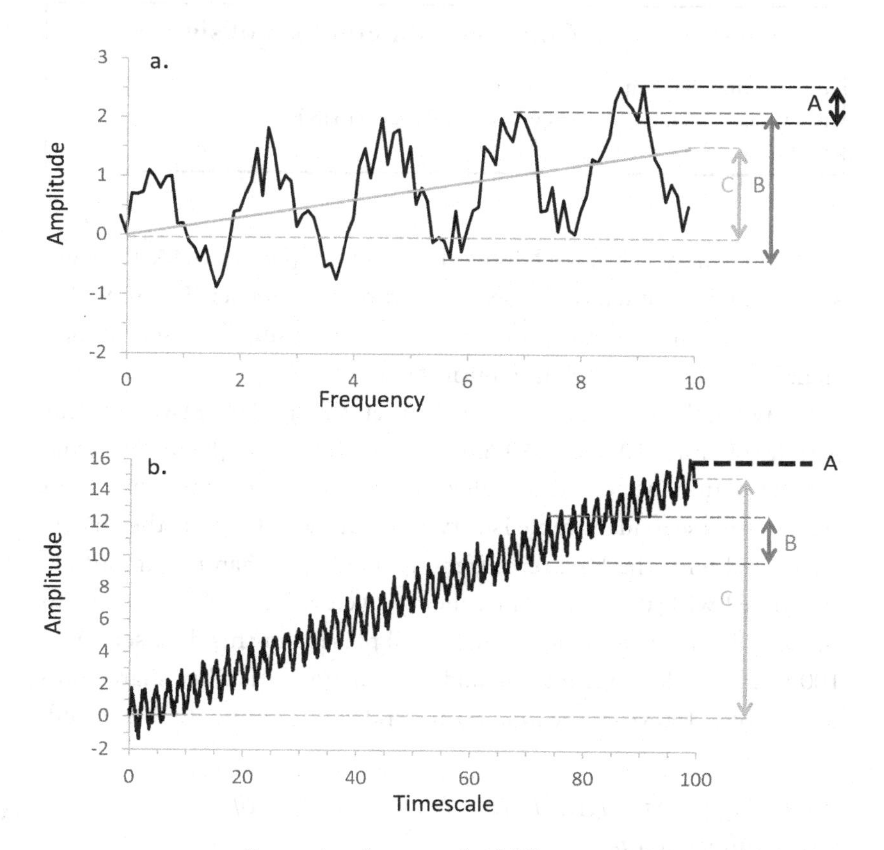

FIGURE 9.3 Comparing the strength of different signals over different timeframes. (a) Data from Example 9.1 plotted over 10 years. (b) Data from Example 9.1 plotted over 100 years. The amplitude of signal A is approximated using black lines. The amplitude of signal B is in approximated using dark grey lines. The change over time associated with signal C is in approximated using light grey lines.

amplitude of the random noise (A) remains ~0.6. The amplitude associated with the linear trend, on the 100-year timeframe, is larger than the amplitudes associated with the noise and medium-frequency signals. In this case, on a 100-year timeframe, the linear trend (C) is the dominant signal.

EXAMPLE 9.2 **Consider a different set of signals**

A. random, high amplitude variability
B. cyclical, medium frequency, low amplitude variability
C. linear trend

The three new fabricated individual signals (Figure 9.4A–C) were summed up to generate the second composite data set (Figure 9.4D), so all three signals are present in the composite data set. Which signal – A, B, or C – is dominant in Figure 9.4D?

When the second composite data set is plotted over two different periods of time, 10 years (Figure 9.4) and 100 years (Figure 9.5), the relative dominance of the signals changes. The slope of the long-term trend in this second example is very low, but over 100 years the change associated with the long-term trend (~5) is larger than the amplitude associated with the medium-frequency cycle (~0.6) and equivalent to the amplitude of the random noise (~5) present in this data set. Over 100 years, the long-term trend and the random variability share dominance; over 150 years the long-term trend would dominate this record.

9.1.1 Signal Strength: Direct Measurements of Atmospheric CO_2

Determining signal strength provides insight into the processes that influence the system over different timescales. The Mauna Loa atmospheric CO_2 data provides a real example of signal dominance changing over time. Atmospheric CO_2 concentrations are influenced both by natural and anthropogenic processes. In Chapter 8 we determined that the annual cycle, with a peak value in May and a minimum value in September / October, exhibits an amplitude of ~6.4 ppm. Over the 58 years of data evaluated, the long-term trend in atmospheric CO_2 concentrations, associated with human burning of fossil fuels, has increased atmospheric CO_2 concentrations by 90.6 ppm with an annual average increase of ~1.56 ppm.

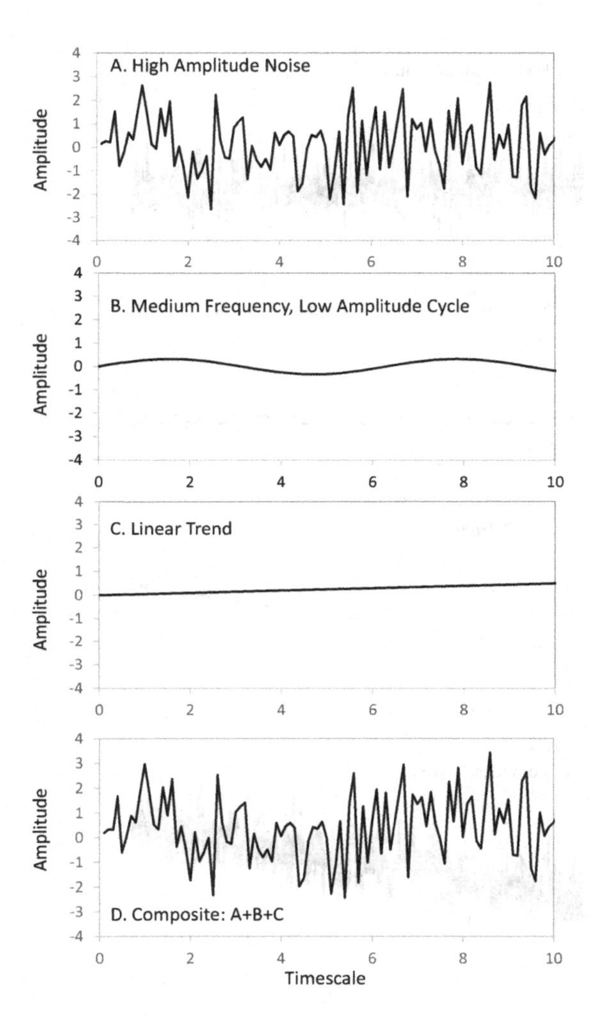

FIGURE 9.4 Example 9.2: Adding multiple signals together to generate a composite data set: 10-year timeframe. The three signals A, B and C were added together to generate D.

When we compare the amplitudes of the two main signals present in the CO_2 data set, it is evident that on a timeframe of one to five years, the annual cycle is the dominant signal. On timeframes longer than five years, the anthropogenic signal becomes the dominant signal. However, the anthropogenic increase in CO_2 is not constant

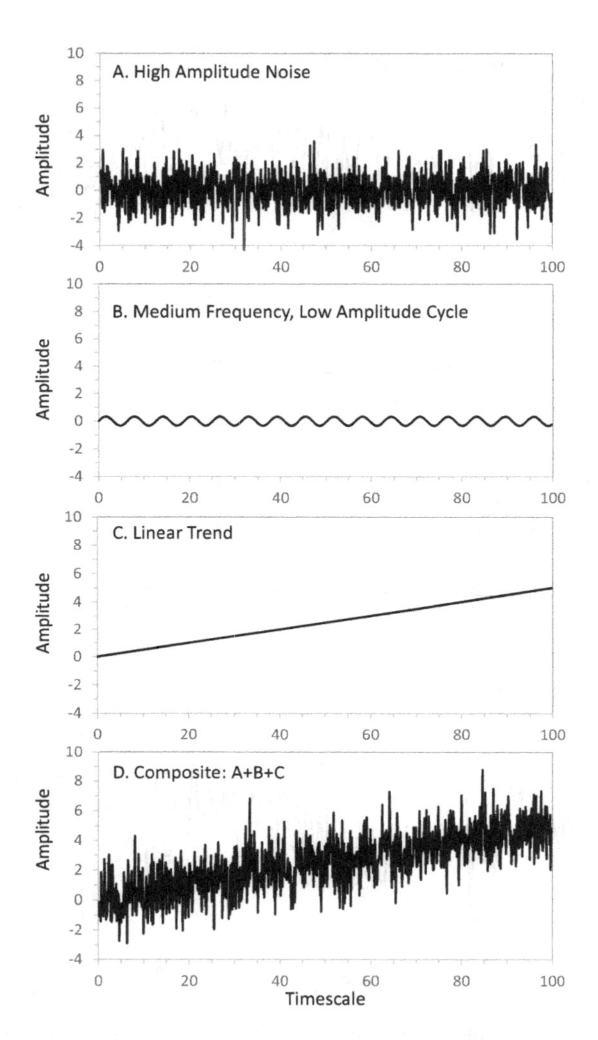

FIGURE 9.5 Adding multiple signals together to generate a composite data set: 100-year timeframe. The same three signals from Figure 9.4 are presented over a longer period of time.

over time. The annual incremental rise in atmospheric CO_2 since 1958 is increasing (Figures 7.4 and 8.2). To highlight the annual change in average CO_2 concentrations, I calculated the interannual difference (the difference between the average CO_2 concentrations for each consecutive year) for the duration of the Mauna Loa data set (Figure 9.6).

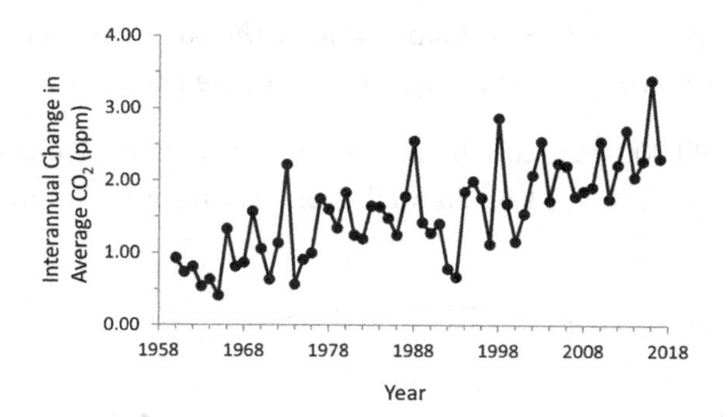

FIGURE 9.6 The interannual change in atmospheric CO_2 from 1958 to 2018 measured at Mauna Loa.

Data from Keeling et al. (2001). http://scrippsco2.ucsd.edu/data/atmospheric_co2/mlo.

- The difference in annual average CO_2 concentrations between 1958 and 1959 was ~0.9 ppm.
- The difference in annual average CO_2 concentrations between 2016 and 2017 was ~2.3 ppm.

Therefore, if this pattern continues, the anthropogenic signal will also become the dominant signal over a timeframe of less than five years.

9.1.2 Signal Dominance: Indirect Measurements of Atmospheric CO_2

The Mauna Loa atmospheric CO_2 record gives us a direct, instrumental, record of CO_2 concentrations in the atmosphere since 1958. But what happened before that?

Was the annual cycle of CO_2 the dominant signal before the industrial revolution, when humans really started using fossils fuels for energy?

How do we get a record of atmospheric CO_2 from a time before we started measuring CO_2 in the atmosphere? From a time before modern humans were even around?

We can use a proxy: ice core data.

Was the annual cycle the dominant cycle in the past, over the 800,000-year timeframe that the ice cores provide us?

No. On the 800,000-year timeframe that the Antarctic glaciers provide, we cannot even see t.he an.nual cycle (Figure 9.7). Taking this

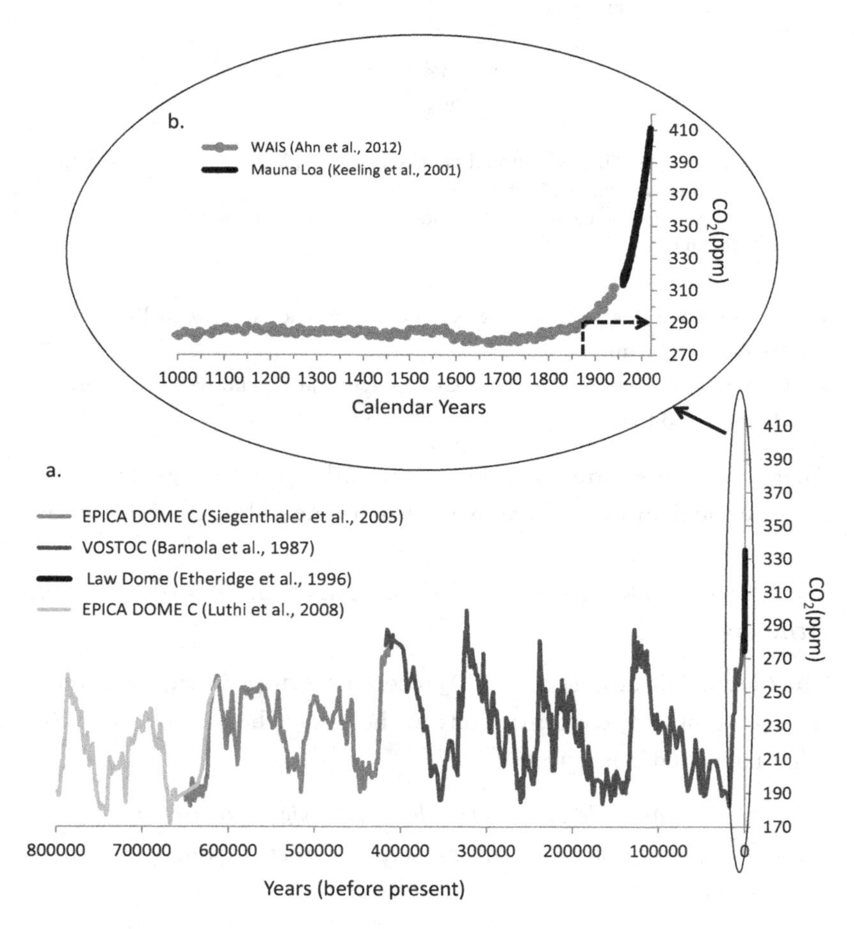

FIGURE 9.7 An 800,000-year Antarctic record of reconstructed atmospheric CO_2 concentrations. (a) Full composite record. (b) 1,000-year long record of atmospheric CO_2 concentrations from both the West Antarctic Ice Sheet reconstruction and direct measurements from the Mauna Loa Observatory in Hawaii. Note: The present-day values are on the right.

long view, another cycle becomes dominant, a cycle that we haven't even noticed using the Mauna Loa CO_2 data. The EPICA / Dome C record shows us that over the last 800,000 years, atmospheric CO_2 has varied with a near 100,000-year cycle.

On a 100,000-year timescale, changes in the Earth's orbit (eccentricity), the tilt of the Earth's axis (obliquity) and the position of the Earth's axis in relation to the Sun (precession) are the processes that control the sources and sinks of atmospheric CO_2. These processes are also related to glaciations. For approximately 800,000 years, atmospheric CO_2 varied between 170 and 290 ppm over a glacial–interglacial cycle, a difference of 120 ppm (Figure 9.7a). The anthropogenic addition of CO_2 to the atmosphere since 1870, the beginning of the industrial revolution, is also about 120 ppm (Figure 9.7b). The atmospheric CO_2 signal associated with anthropogenic activity is now equivalent to the CO_2 change associated with past glacial cycles.

9.2 SIGNALS VS. NOISE

If there are multiple signals of variability present in the data, can we always isolate them?

When multiple non-interacting cycles are present in the same data set, they can be systematically isolated by removing the highest frequency cycle, through averaging, and then subtracting the average from the raw data. If more than two independent cycles coexist in a data set, this process can be repeated.

For a composite data set with three non-interacting signals (A, B, and C), isolate the individual signals by

1. finding an appropriate averaging window to remove the highest frequency signal (A);
2. using a running average to remove the highest frequency signal (A) from the original composite data set (A + B + C);
3. subtracting the running average (B + C) from the original composite data set (A + B + C) to isolate the random noise (A);

4. finding a new averaging window to remove the medium frequency signal (B);
5. using the new running average to remove the medium-frequency signal (B) from the remaining composite data set (B + C) and isolate the long-term trend (C); and
6. subtracting the long-term trend (C) from the remaining composite data set (B + C) to isolate the medium frequency signal (B).

In the two fabricated examples above, I generated the composite data sets by adding together three signals with different frequencies and amplitudes. The first example was built using a synthetic data set with the following three signals:

A. random low amplitude noise
B. cyclical medium frequency, medium amplitude variability
C. linear trend

In order to retrieve these same signals from the composite data set, I will follow the procedure outlined above. For this example, I first chose an averaging window of eight data points. The random noise has no obvious cycle. Also, the amplitude associated with the random noise is low and equally distributed about 0. So, averaging over eight datapoints is sufficient to remove most of the low amplitude noise (Figure 9.8a). I will call the new averaged data set B + C. Subtracting B + C from the original data set will isolate the random noise (A) (Figure 9.8b).

To retrieve the medium-frequency variability, I used a running average again. This time I averaged B + C with the aim of finding the long-term trend (C) (Figure 9.8c). I used an averaging window of two years. If you count, you will find that the medium-frequency cycle oscillates 10 times every 20 years (or once every 2 years). Therefore, I chose an averaging window equivalent to the frequency of the medium-scale cycle.

There is one final step. To retain the medium-frequency cycle (B) in isolation, I have to remove the long-term trend. I did this by subtracting the long-term trend (C) from the composite of the

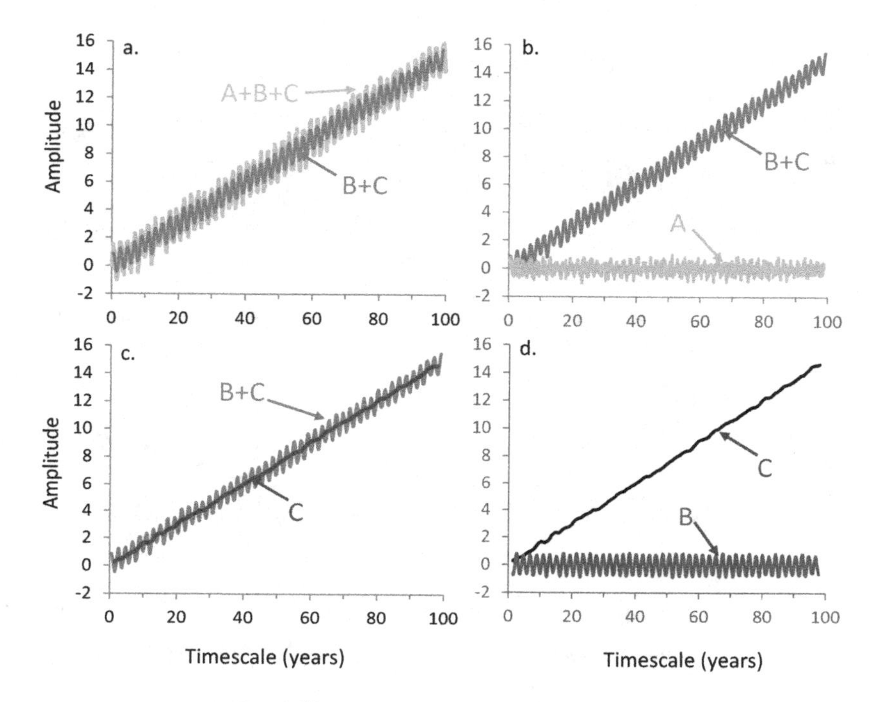

FIGURE 9.8 Example 9.1: Deconstructing a composite data set comprising three signals A, B, and C. (a) A running average applied to the composite data set (A + B + C) generates a new data set with A removed. (b) Subtracting the new data set (B + C) from the original (A + B + C) retrieves signal A alone. (c) The process is repeated. Averaging B + C will remove B. (d) Subtracting C from B + C will isolate B.

medium-frequency cycle and the long-term trend (B + C). You can see in Figure 9.8d that by subtracting the long-term trend (C) from B + C, I can, in fact, isolate a signal very close the original medium-frequency cycle.

The second example was built from the following signals:

A. random, high amplitude variability
B. cyclical, medium frequency, low amplitude variability
C. linear trend

In Example 9.2, the amplitude of the random noise is large compared to the other two signals. The averaging window of 10 datapoints reduces

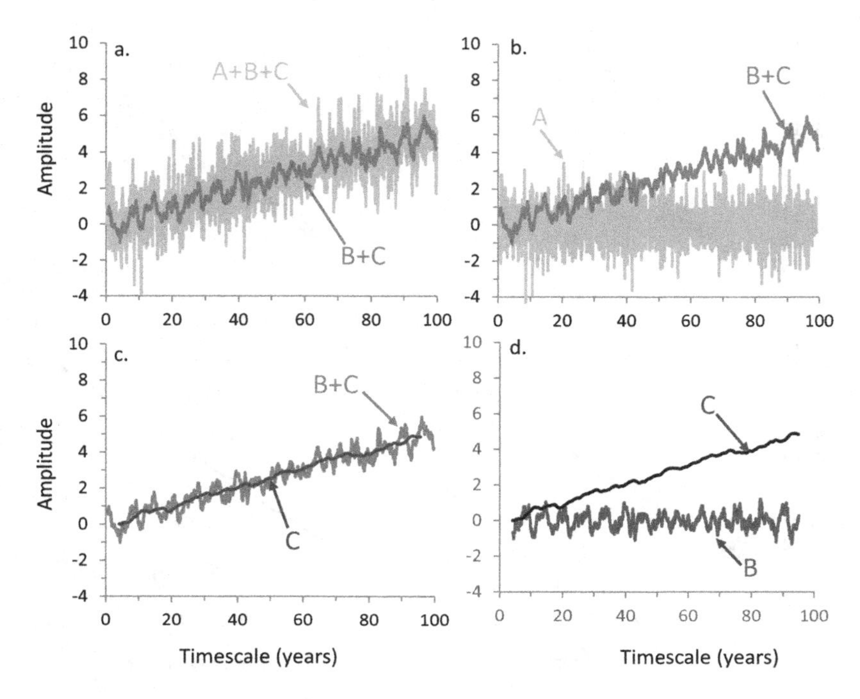

FIGURE 9.9 Example 9.2: Deconstructing a composite data set comprising three signals A, B, and C. Isolating individual signals can be difficult if the timescales and amplitudes are too similar. (a) A running average applied to the composite data set (A + B + C) generates a new data set with A removed. (b) Subtracting the new data set (B + C) from the original (A + B + C) retrieves signal A alone. (c) The process is repeated. Averaging B + C will remove B. (d) Subtracting C from B + C will isolate B.

some of the noise, but the isolated medium-frequency signal (B) does not closely resemble the original medium-frequency cycle. Even when the long-term trend has been removed, the original medium-frequency variability that was used to build the composite data set cannot be fully retrieved (Figure 9.9).

Why?

In Example 9.2, the signal to noise ratio is low. Averaging the original random noise signal used for Example 9.2 generates a signal that has a

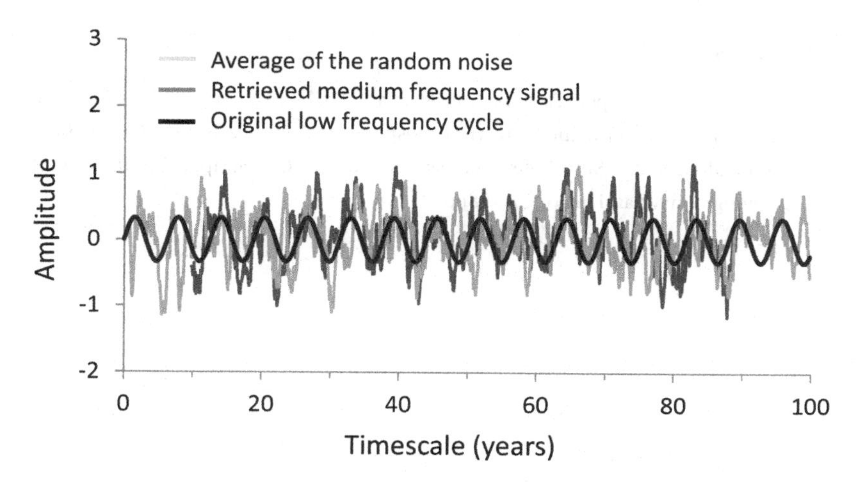

FIGURE 9.10 Comparing the medium-frequency variability in Example 9.2. The original medium-frequency variability (black) cannot be retrieved. The retrieved medium-frequency signal (dark grey) is highly influenced by the random noise (light grey).

timescale similar to the original medium-frequency signal (Figure 9.10). The random noise in Example 9.2 is interfering with signal B, making it impossible to cleanly retrieve the medium-frequency signal. Even more technical approaches to timeseries analysis would have difficulties retrieving the original medium-frequency signal used in Example 9.2. Following the five steps outlined earlier, the long-term trend can be removed from the data set, but the original medium-frequency variability cannot be cleanly retrieved.

9.3 TAKE-HOME MESSAGES

- The relative amplitude determines the dominance of the signals.
- The dominant signal can change depending on the timeframe considered.
- Determining signal dominance provides insight into the relative importance of different processes influencing the system over different timescales.
- It is not always possible to isolate all the signals, even if you know they are present in a data set.

REFERENCE

Keeling, C. D., S. C. Piper, R. B. Bacastow et al. (2001). Exchanges of atmospheric CO_2 and $13CO_2$ with the terrestrial biosphere and oceans from 1978 to 2000. I. Global aspects, SIO Reference Series, No. 01-06, Scripps Institution of Oceanography, San Diego, CA, 88pp.

Chapter 10

Preparation

1. Before reading Chapter 10, consider the following questions:
   - How does the mean value of your data set relate to the distribution of your data?
   - What is a standard deviation? Why is it useful?
2. How does the lifespan of an individual organism relate to the average lifespan of a group of this same type of organism?
3. Draw a schematic diagram or graph that reflects the answer that you gave in question (2) above.

10 Characterizing Your Data

10.1 CHARACTERIZING YOUR DATA

Environmental data is more than a bunch of numbers. Meaningful information about the natural world is embedded in those numbers as patterns, cycles, trends, changes, and events. Arranging your data in different ways can highlight different features of the phenomena captured by your data set. This dual aspect of a data set is its power: data is quantifiable information. You can describe the key features of your data set using words, but you can also quantify, or characterize, critical information using numbers, mathematical equations, or statistical concepts.

10.1.1 Mean, Median, and Mode

The **mean,** or average value, of a data set is found by summing all of the values in the data set and dividing by the number of values in the data set. The **median,** or middle value, of the data set is found by counting the number of values in the data set and choosing the middle value. If the number of values in the data set is even, the median is calculated by taking the average value of the middle two datapoints. The **mode,** or most common value, is found by identifying repeat values and selecting the value that is repeated the most (Table 10.1). Depending on the distribution of the data, the central tendency (the centre position or location) of the data might best be described by the mean, median, or the mode.

Data that is **normally distributed** has a symmetrical probability distribution with mean, median, and mode values that all coincide (Figure 10.1). We call the shape of normally distributed data a bell

Table 10.1 *Mean, median, and mode*

Mean	Median	Mode
2	2	2
2	2	2
2	2	2
3	3	3
3	3	3
3	3	3
3	3	3
4	4	4
4	4	4
5	5	5
5	5	5
6	6	6
6	6	6
7	7	7
7	7	7
8	8	8
8	8	8
9	9	9
9	9	9
5	**5**	**3**

The calculated value for each parameter is in the bottom cell of each row. Mean is calculated as the average of all data. Median is the middle datapoint in the series. Mode is the most common value.

curve. The mean and the variability of the data around the mean define the shape of the bell curve for the phenomenon of interest.

Many environmental phenomena, however, produce data sets that are not normally distributed. Real data sets might be negatively skewed (long tail on the left) or positively skewed (long tail on the right), be exponentially distributed, or have more than one peak value (Figure 10.1).

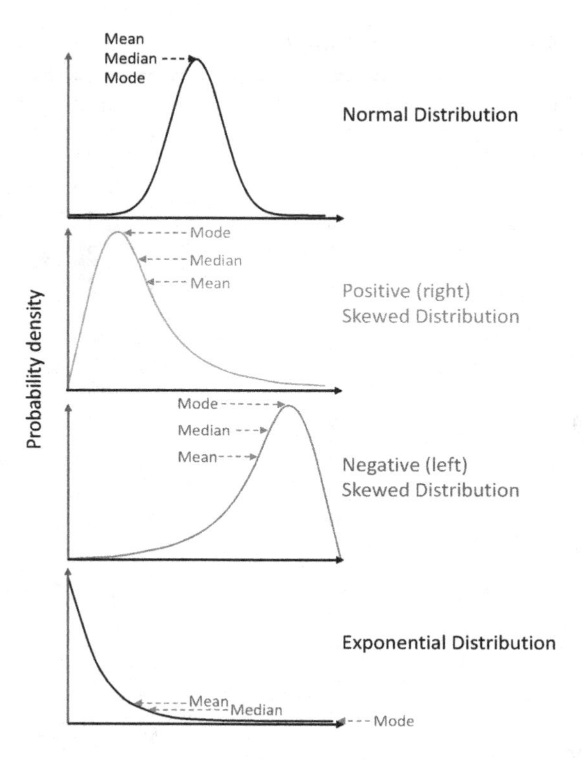

FIGURE 10.1 A cartoon depiction of normally distributed data (in black) and skewed data (in grey).

10.1.2 Data Examples: The Fraser River and Fraser River Sockeye Salmon

The Fraser River, on the south coast of British Columbia, Canada, is a critical spawning ground for Pacific Salmon. Fed by melting mountain snowpack and seasonal rainfall, the Fraser River discharge is highest in the spring and summer and lowest in the winter. Sockeye salmon remain in the Fraser River for one to two years after hatching and then migrate to the Pacific Ocean. The sockeye return after two to three years in the ocean to the same location where they were born. Salmon returns show a strong four-year cycle, but overall the sockeye salmon returns are declining. The success of the return is influenced by both riverine and oceanic conditions throughout their lives.

10.1.2.1 Fraser River Water Level and Discharge

Both the water level (m) and discharge (m^3/s) of the Fraser River are monitored by the Canadian Government and the data is publicly available.[1] At a site near Hope, British Columbia (BC), the Fraser River water level was measured near continually between 1912 and 2016 producing data with

- mean = 4.873
- median = 4.465
- mode = 3.579

When the water level data is plotted as a time series, you can see that the central tendency (the most dense data) varies slightly over time and that the range of values above the central tendency is larger than the range of values below the central tendency (Figure 10.2a).

Focusing on one year only (Figure 10.2b) highlights the annual structure that is present in the data. There is a dominant peak in the summer (reflecting the snowmelt) and a smaller peak in the fall (reflecting seasonal rain). When plotted as a histogram, it is evident that the water-level data is positively (right) skewed (Figure 10.3): the three measurements of central tendency (mean, median, and mode) are not equal, the shape of the distribution curve is not symmetrical, and the long tail of the distribution is on the right of the peak. The presence of the plateau shoulder suggests that two separate processes are acting together, both influencing the water height of the Fraser River. Together the time series and histogram illuminate that the underlying phenomena of Fraser River water level has

[1] Data retrieved from https://wateroffice.ec.gc.ca/search/historical_e.html.

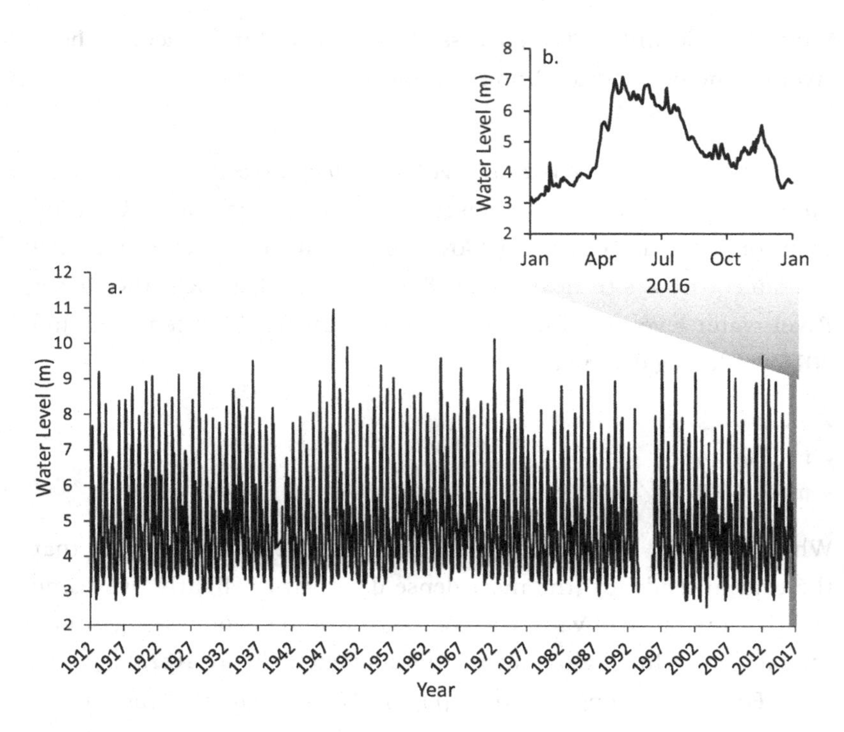

FIGURE 10.2 A time series of annual water level (m) of the Fraser River, BC, Canada, retrieved from the Government of Canada (https://wateroffice.ec.gc.ca/search/historical_e.html). (a) Fraser River water level from 1912 to 2017. (b) Monthly Fraser River water level for 2016.

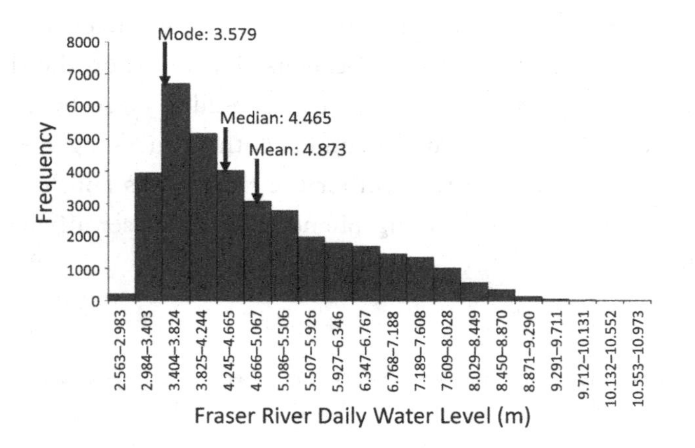

FIGURE 10.3 The distribution of the annual water level (m) of the Fraser River, BC, Canada, retrieved from the Government of Canada (https://wateroffice.ec.gc.ca/search/historical_e.html).

- a central tendency that varies slightly (between 4 and 5) annually;
- periodic increases in water level that are generally more intense than the periodic decreases in water level; and
- at least two separate processes with influence: likely snowmelt dominating in the spring and summer and rainfall dominating in the fall.

The Fraser River discharge data from the same location – Hope, British Columbia – looks similar to the water-level data when plotted as a time series (Figure 10.4) but looks different when plotted as a histogram (Figure 10.5). Again, the mean (2,721), median (1,880), and mode (1,010) values for this data are not all the same, so we know that the data are not normally distributed. The histogram of the water discharge shows that the data are exponentially distributed: many relatively small values and fewer relatively large values. The same shoulder plateau is present in the discharge data that we noticed in the water height data. This slightly changes the shape of the exponential distribution.

The different distributions of the water level and discharge data sets imply that different processes are underlying the data collected. Of course, because this is real data, there could also be error involved. To determine if the differences between these data sets are real and significant, more complex statistics would need to be employed.

10.1.2.2 Fraser River Sockeye Salmon

In the Northeast Pacific, the local commercial and recreational salmon fisheries take place in the ocean, catching adult salmon en route to spawn in their native rivers. Escapement, the number of salmon that escape the fishery and reach the river, is one measure used to predict the size of the annual sockeye cohort. Data representing the 2017 sockeye salmon escapement[2] shows that on July 1, sockeye were only showing up in small numbers in the Fraser River (Figure 10.6). The number of salmon returning to the Fraser River

[2] Krivanek and Whitehouse (2017).

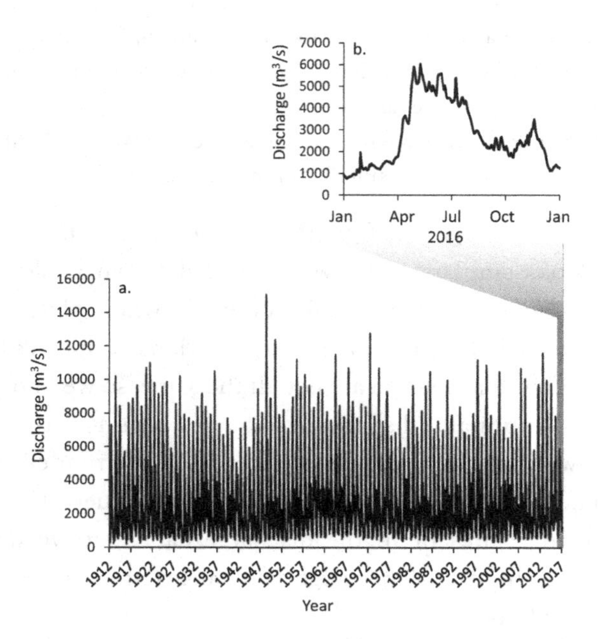

FIGURE 10.4 A time series of annual water discharge in m^3/s of the Fraser River, BC, Canada, retrieved from the Government of Canada (https://wateroffice.ec.gc.ca/search/historical_e.html). (a) Fraser River discharge from 1912 to 2017. (b) Monthly Fraser River discharge for 2016.

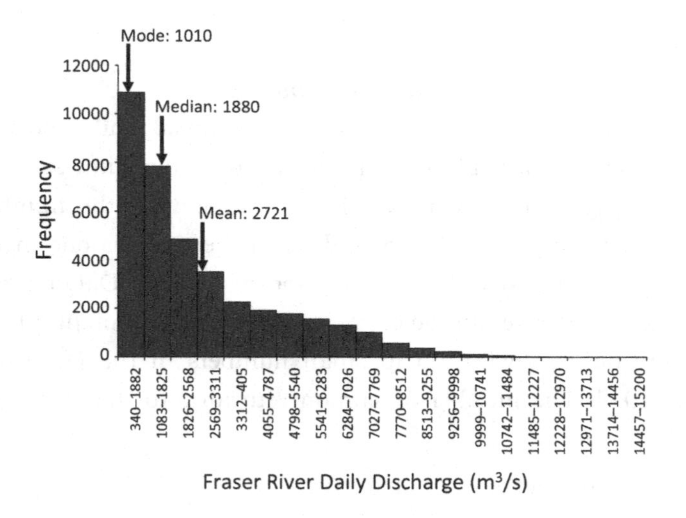

FIGURE 10.5 The distribution of the annual water discharge in m^3/s of the Fraser River, BC, Canada, retrieved from the Government of Canada (https://wateroffice.ec.gc.ca/search/historical_e.html).

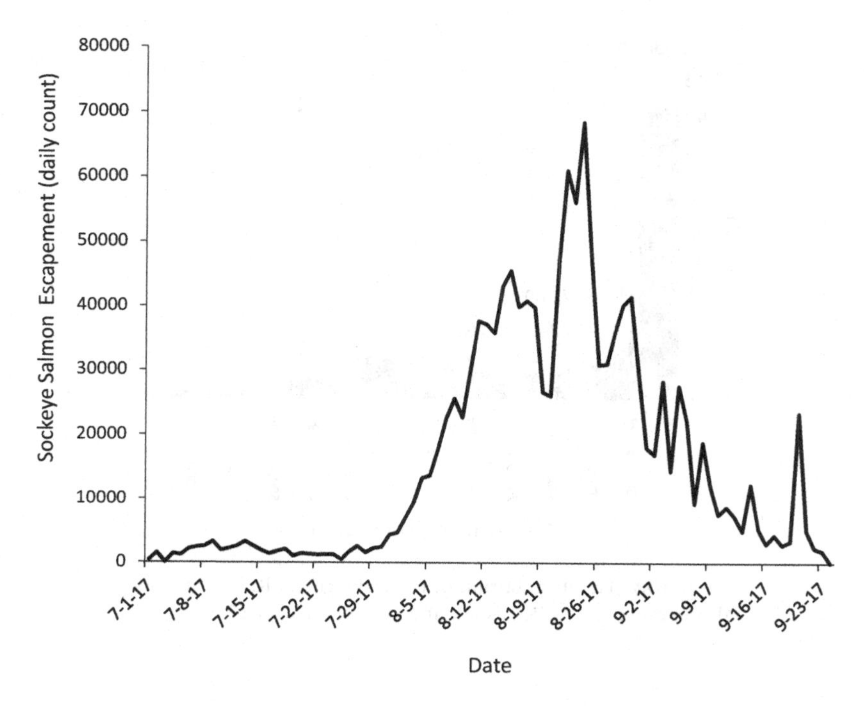

FIGURE 10.6 A time series of the 2017 Fraser River daily sockeye salmon escapement (Krivanek and Whitehouse, 2017).

increased through August with the peak of the 2017 return occurring on August 24. The 2017 sockeye escapement data is not normally distributed: the histogram shows a peak frequency in the first bin (0–5,000) and then a relatively uniform distribution (Figure 10.7). There are only three days with the highest escapement values. This distribution reflects the underlying process:

- The initial salmon run through July was slow, with 0 sockeye returning on most days.
- Through August, between 20,000–40,000 sockeye were regularly showing up in the river.
- There were only three days when the number of returning salmon was over 56,000.

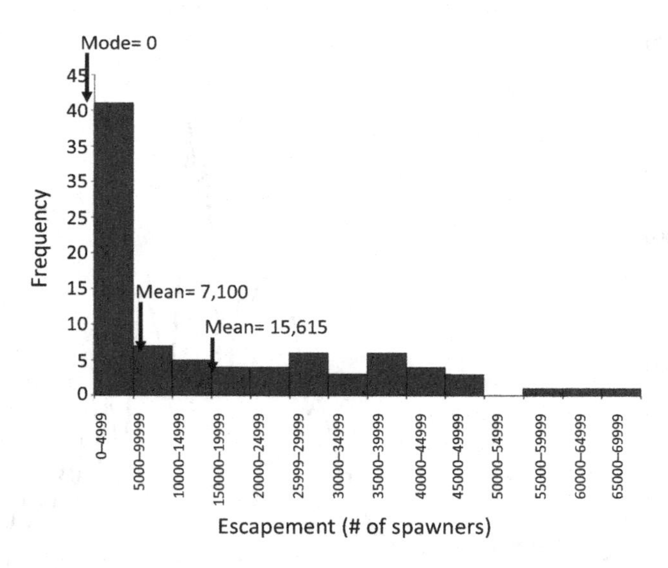

FIGURE 10.7 The distribution of the 2017 Fraser River daily sockeye salmon escapement (Krivanek and Whitehouse, 2017).

The mean (15,615), the median (7,100), and the mode (0) would all have limitations if used to describe the central tendency of this data.

10.1.3 Standard Deviation and Variance

The mean, median, and mode can all be used to provide different information about the central tendency of a data set. Additionally, the **range** of the data provides information about the highest and lowest values of the data set, but it doesn't tell you about distribution of the data in between. Knowing how the data cluster about the mean allows for a deeper understanding of the data set and the underlying phenomena.

The **standard deviation** (σ) is a mathematical parameter that provides information about the distribution of the data around the mean value. For normally distributed data, the standard deviation provides information about the width of the bell curve, and therefore an intuitive understanding of how tightly the data are clustered about the mean. For a normal distribution, one, two, and three standard

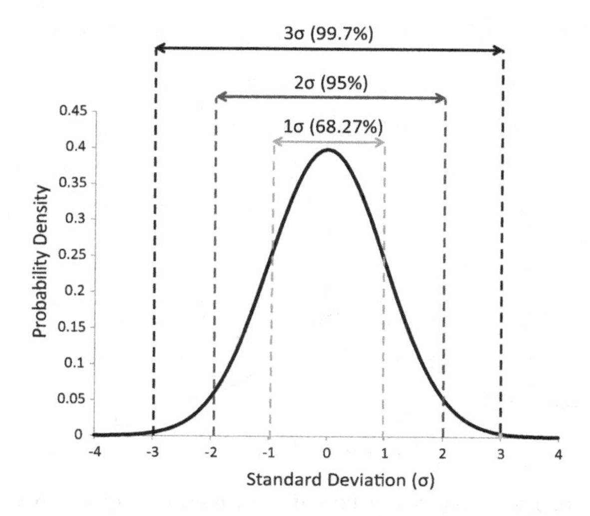

FIGURE 10.8 A cartoon depiction of 1, 2, and 3 standard deviations (σ) for a normally distributed data set.

deviations are used to describe ranges around the mean within which a prescribed proportion of the data points fall (half will be above the mean and half will be below) (Figure 10.8):

- One standard deviation (1σ) captures 68.27% of the data points
- Two standard deviations (2σ) capture 95% of the data points
- Three standard deviations (3σ) capture 99.7% of the data.

I have plotted two different theoretical normal distributions (Figure 10.9). A small standard deviation indicates that the data are very tightly distributed around the mean: the resultant bell curve is tall and skinny indicating that there is little variation in the underlying phenomenon. Plot A is an example of a small standard deviation. A large standard deviation indicates that the data are widely distributed around the mean: the resultant bell curve is low and broad, representing a phenomenon that is quite variable. Plot B is an example of a larger standard deviation.

Variance (σ^2) is another tool that can be used to describe the distribution of data around the mean. It is important to recognize whether the standard deviation or the variance is being reported.

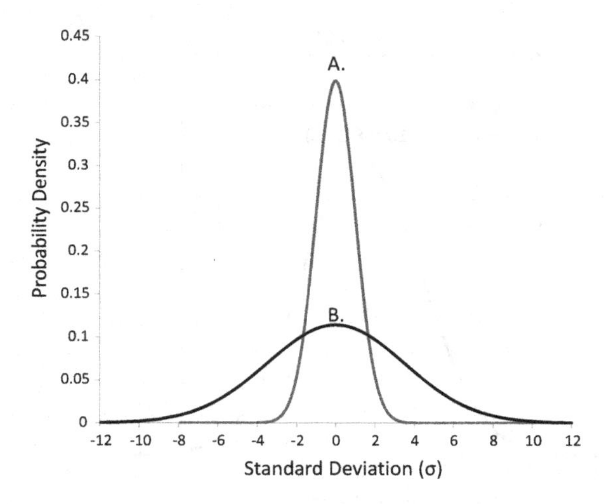

FIGURE 10.9 Two hypothetical normal distributions with different means and standard deviations (σ). Curve A: mean = 0, σ = 1. Curve B: mean = 0, σ = 3.5.

They are similar, but not identical. The standard deviation (σ) is reported in the same units as the original data set, whereas that variance (σ^2) is the standard deviation squared. It is, therefore, easier to interpret the standard deviation in relation to the original data set than it is to interpret the variance.

The sample variance is calculated using the equation

$$\text{Sample variance } (\sigma^2) = \frac{\sum (x_i - x_{avg})^2}{(n-1)}$$

where

- x_i are the individual x-values
- x_{avg} is the average of the x-values
- n is the total number of x-values

The standard deviation is calculated by taking the square root of the variance:

$$\text{Standard deviation } (\sigma) = \sqrt{\sigma^2}$$

Calculating the standard deviation is quite straightforward in a spreadsheet. Create individual columns for each of the steps to keep track of the calculations. Using fabricated data as an example, the standard deviation calculation is demonstrated in Table 10.2. The upper panel (Table 10.2a) shows the data and the results of the calculations. The lower panel (Table 10.2b) shows the Excel equations that were used to generate the calculations.

Interpreting the variance and standard deviation for non-normal distributions is not intuitive.

10.2 DESCRIBING THE MONTHLY ATMOSPHERIC CO_2 DATA USING AVERAGES AND STANDARD DEVIATIONS

At the end of Chapter 8, after a preliminary investigation of the Mauna Loa atmospheric CO_2 data, three ways to investigate variability in the annual cycle of CO_2 in Hawaii were presented:

- Investigate the monthly variability over the timeframe of the full data set.
- Investigate whether there has been a change in the amplitude of the annual cycle over time from 1959 to 2017.
- Determine whether there has been a unidirectional trend in maximum and minimum CO_2 values observable since 1959.

The first approach can be undertaken using the monthly mean and standard deviation calculations of the Mauna Loa CO_2 data set because we have a large data set to work with that, when presented to show monthly variability, is essentially normally distributed.

The annual cycle of CO_2 variability is driven by the natural seasonal cycle of photosynthesis and respiration. In Chapter 8, we decomposed the raw CO_2 data into two constituent components: the seasonal cycle of atmospheric CO_2 and the long-term trend. With the use of a 12-month running mean, the anthropogenic component of the atmospheric CO_2 signal was isolated. Subtracting the 12-month running mean from the raw data then isolated the seasonal cycle as an anomaly data set.

Table 10.2 *Example spreadsheet demonstrating how to calculate standard deviation*

(a) Standard deviation calculations using fabricated data

	A	B	C	D	E
1	**Sample**	**Value**	**Average**	**Difference between the sample value and the average**	**Squared difference between the sample value and the average**
2	1	45	48.6	3.6	12.96
3	1	35	48.6	13.6	184.96
4	1	75	48.6	−26.4	696.96
5	1	54	48.6	−5.4	29.16
6	1	34	48.6	14.6	213.16
7	1	65	48.6	−16.4	268.96
8	1	23	48.6	25.6	655.36
9	1	76	48.6	−27.4	750.76
10	1	45	48.6	3.6	12.96
11	1	34	48.6	14.6	213.16
12					
13	1	67.0	**= + 1 std dev.**	**Total # of samples =**	10
14	1	30.2	**= −1 std. dev.**	**Degrees of freedom =**	9
15				**Variance =**	337.6
16				**Std. Dev.=**	18.4

Table 10.2 (*cont.*)

(b) The equations used to calculate the values in (a)

	A	B	C	D	E
1	**Sample**	**Value**	**Average**	**Difference between the sample value and the average**	**Squared difference between the sample value and the average**
2	1	45	=AVERAGE (B2:B11)	=C2-B2	=D2^2
3	1	35	=$C$2	=C3-B3	=D3^2
4	1	75	=$C$2	=C4-B4	=D4^2
5	1	54	=$C$2	=C5-B5	=D5^2
6	1	34	=$C$2	=C6-B6	=D6^2
7	1	65	=$C$2	=C7-B7	=D7^2
8	1	23	=$C$2	=C8-B8	=D8^2
9	1	76	=$C$2	=C9-B9	=D9^2
10	1	45	=$C$2	=C10-B10	=D10^2
11	1	34	=$C$2	=C11-B11	=D11^2
12					
13	1	=C2 +E16	**= + 1 std dev.**	**Total # of samples =**	=COUNT(A2: A11)
14	1	=C2- E16	**= - 1 std. dev.**	**Degrees of freedom =**	=E13–1
15				**Variance =**	=SUM(E2: E11)/E14
16				**Standard deviation =**	=SQRT(E15)

Using this anomaly data set, we can focus on the variability within each month over the timeframe of the full data set: we can determine the range, mean, and standard deviation of all January values, all February values, and so on within this data set.

To characterize the monthly variability, the CO_2 anomaly data must be reorganized by month. (Currently it is organized by year.) Once the monthly data is collated, the distribution can be checked to determine if the data is normally distributed or not. Then, the mean and standard deviations for each month can be calculated and plotted to highlight the monthly variability over the entire data set.

10.2.1 Step 1: Checking the Distribution of the Data

For a review of the basics of working in Excel, see the Appendix. Depending on the version of Excel that you have, you might be able to choose "histogram" as a chart option. If that option is not available to you, then you can make your own. Within your Excel spreadsheet, highlight all of the data. Then

- Sort the data using the column representing months so that all of the January values (labelled 1 in the month column) and February values (labelled as 2 in the same column), and so on, are collected together. All of the other data must be highlighted to ensure that the individual cells remain associated with the correct date.
- Copy and paste the data for only one month into a new spreadsheet.
- On the new spreadsheet, sort the data to order the CO_2 anomalies from lowest to highest values (Table 10.2).
- Choose a bin size that will give you at least five bins.
- Make two new columns, one for the bins and one for the frequency.
- Count the number of values that fall into a bin, using the command
  =count(cell range)
  and fill the frequency column with this information.
- Plot a histogram with the bins on the *x*-axis and the frequency of values on the *y*-axis.

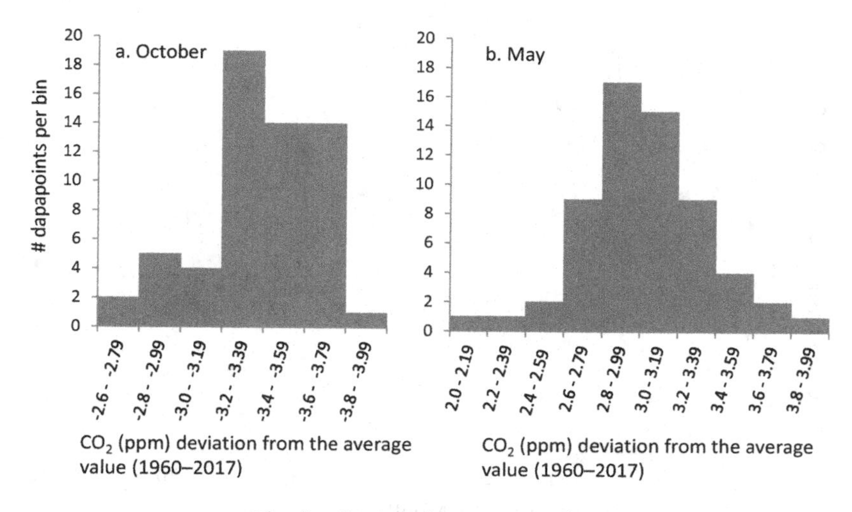

FIGURE 10.10 The distribution of average October and May CO_2 residuals. (a) October residuals. (b) May residuals.
Data from Keeling et al. (2001).

I plotted histograms for May and October, the two months that I plan to work with (Figure 10.10). Though the histograms for these two months are different, both display general bell-like shapes. The histogram for May is a classic bell curve: mean = 3; median = 3. The October values are higher than those from May: mean = 3.4; median = 3.4. The histogram for October is taller and tighter.

10.2.2 Step 2: Generating Monthly Averages

Back on your CO_2 anomaly spreadsheet that is sorted by month:

- Remove the incomplete years (1958/1959) at the beginning of the data set. This will allow each average to be made from the same number of values.
- Calculate an average value for each individual month.
- Plot the average monthly anomaly values from 1960 to 2017 (Figure 10.11).

10.2.3 Step 3: Calculating Monthly Standard Deviations

Make a new spreadsheet, based on Table 2.1, to calculate the standard deviation of the anomaly values for each month:

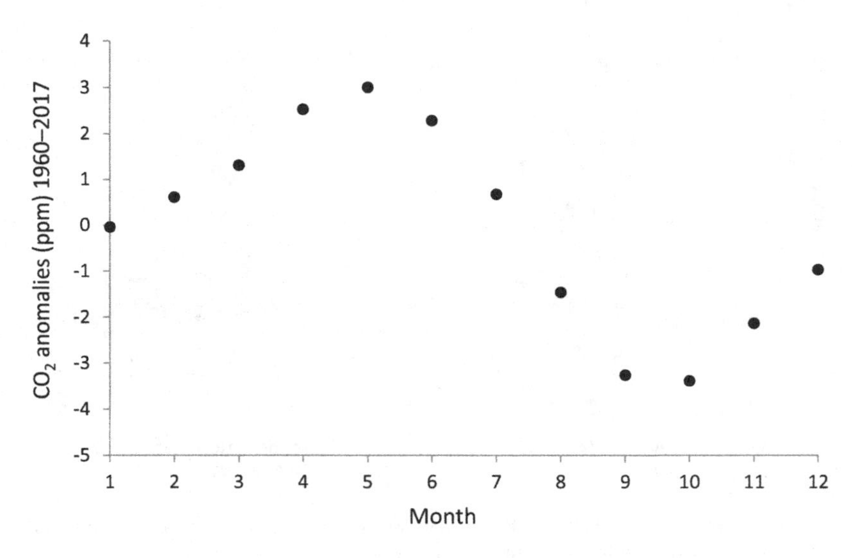

FIGURE 10.11 Monthly CO_2 anomalies from 1960 to 2017.
Data from Keeling et al. (2001).

- Calculate two standard deviations (2σ).
- Add customized error bars to your graph showing two standard deviations (2σ) (Figure 10.12).

Plotting 2σ on the graph means that 95% of the data will fall between the upper and lower bounds of the error bars. When plotting data, you can choose whether to represent one, two, or three standard deviations. You have to make sure you clearly communicate what is presented. Figure 10.12 shows that the inter-annual variability in atmospheric CO_2 concentrations for each month is not the same: April (2σ = 0.8), August (2σ = 0.7), and September (2σ = 0.9) vary the most from year to year, while November (2σ = 0.4) and December (2σ = 0.4) vary the least from year to year.

10.2.4 Summarizing the Results

After performing this monthly analysis of atmospheric CO_2 for 57 years of data (1960–2017) collected from Mauna Loa observatory, we can

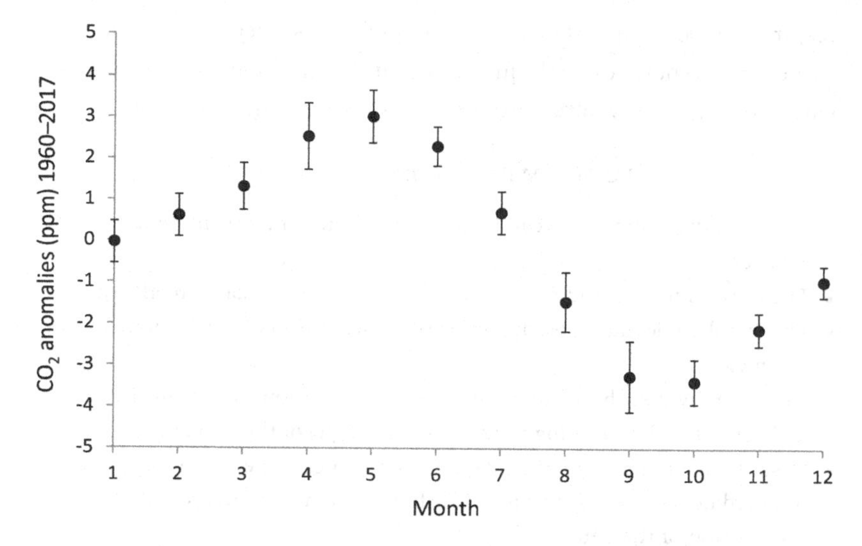

FIGURE 10.12 Monthly CO_2 anomalies and standard deviations from 1960 to 2017.
Data from Keeling et al. (2001).

confirm that there is inter-annual variability in the monthly values over this timeframe. The standard deviations range from $2\sigma = 0.4$ in December to $2\sigma = 0.9$ in September with the lowest standard deviations found in the winter months (November $2\sigma = 0.4$, December $2\sigma = 0.4$, January $2\sigma = 0.5$, and February $2\sigma = 0.5$) and highest standard deviations found in the spring (April $2\sigma = 0.8$) and late summer (August $2\sigma = 0.7$, September $2\sigma = 0.9$).

Though this analysis is useful in determining variability in the monthly average CO_2 concentrations, we currently have no understanding if there is a pattern associated with the inter-annual variability. We now know that the variability in non-anthropogenic atmospheric CO_2 concentrations in the spring (the end of the respiration season) and the late summer (the end of the photosynthetic season) are higher than the variability of the winter months. This might indicate that between 1960 and 2017 the growing season has intensified, with more photosynthesis occurring through the summer, and consequently more

respiration occurring though the winter. This speculation could be turned into a new research question that would drive a deeper investigation into monthly variability of atmospheric CO_2.

10.3 TAKE-HOME MESSAGES

- Mean, median, and mode can be used to describe the central tendency of a data set.
- The mean, median, and mode of normally distributed data are all equal.
- The standard deviation (σ) describes the distribution of data around a mean value.
- For normally distributed data, the standard deviation can be intuitively understood as determining the width and height of the bell curve.
- Describing, or characterizing, data using the mean, median, mode, and standard deviation is a way to quantify important information about the distribution of the data.

REFERENCES

Keeling, C. D., S. C. Piper, R. B. Bacastow et al. (2001). Exchanges of atmospheric CO_2 and $13CO_2$ with the terrestrial biosphere and oceans from 1978 to 2000. I. Global aspects, SIO Reference Series, No. 01-06, Scripps Institution of Oceanography, San Diego, CA, 88pp.

Krivanek, J. and T. Whitehouse (2017). 2017 In-season Escapement Estimates of Fraser River Salmon at Qualark Dual Frequency Identification Sonar (DIDSON) Site with Test Fishing Results and Species Apportionment. 2017 Project Report to the Southern Boundary Restoration and Enhancement Fund. Fisheries and Oceans Canada. Available at www.psc.org/

Chapter 11

Preparation

1. Before reading Chapter 11, consider the following questions:
   - Can one data set, from one location, really represent a global phenomenon?
   - Do other sites show the same signal?
   - How can I compare data collected from two (or more) different locations?
2. Consider the following graphs. Identify which one(s) show a strong common variability between the two data sets and which one(s) show weak common variability between the two data sets?

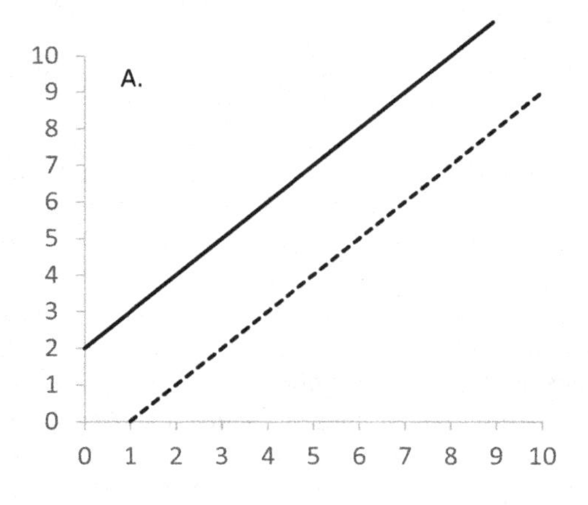
A.
10
9
8
7
6
5
4
3
2
1
0
0 1 2 3 4 5 6 7 8 9 10

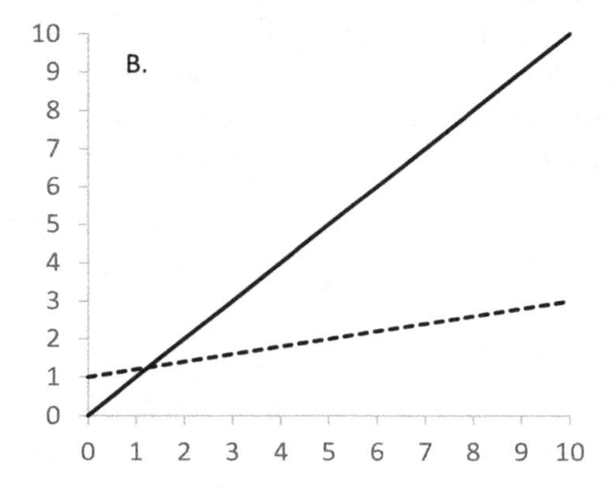
B.
10
9
8
7
6
5
4
3
2
1
0
0 1 2 3 4 5 6 7 8 9 10

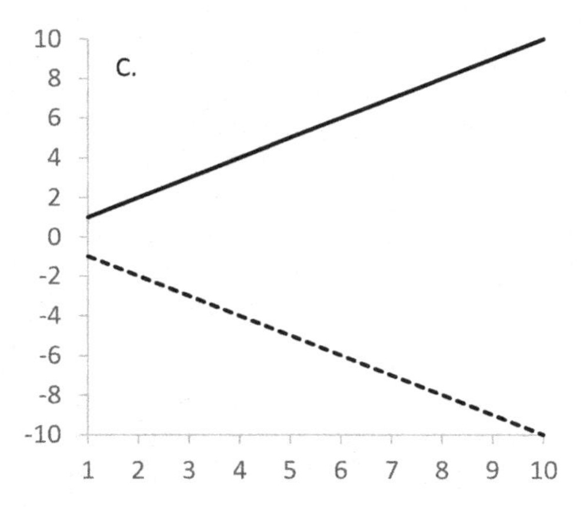
C.
10
8
6
4
2
0
-2
-4
-6
-8
-10
1 2 3 4 5 6 7 8 9 10

11 Comparing Data Sets

11.1 PLACING A LOCATION IN CONTEXT

We have used the Mauna Loa monthly CO_2 data set to isolate and characterize changes in the anthropogenic and biological contributions to atmospheric CO_2 concentrations between 1960 and 2017. We have also studied variability in the seasonal cycle, driven by changes in photosynthesis and respiration from year to year. *But is the CO_2 record from Mauna Loa, a mountain in Hawaii, representative of the global variability in atmospheric CO_2?*

As you become more familiar with the data set you are using, you might want to compare it with data collected from other locations to get a deeper understanding of your phenomenon of interest or to gain context for your particular data set. You might ask:

Are the seasonal cycle and long-term trend evident in atmospheric CO_2 concentrations collected from Mauna Loa, Hawaii, representative of local or global phenomena?

To answer this question, you have to find some comparative data from different locations and develop a new technique: correlation analysis.

11.1.1 Choosing Comparative Data Sets

Atmospheric CO_2 has been monitored in a number of locations by a number of different agencies. In addition to the Mauna Loa Observatory, the Scripps CO_2 program[1] has been collecting data from the South Pole and Point Barrow Alaska since 1957 and 1967, respectively. These two

[1] http://scrippsco2.ucsd.edu/data/atmospheric_co2/sampling_stations

additional locations allow a good comparison between the tropics (Mauna Loa), the high Northern Hemisphere (Point Barrow, Alaska), and the high Southern Hemisphere (South Pole).

There are different data options of data available from each site: daily and monthly, in situ and flask samples. *Which data sets should we choose?*

We will be comparing the atmospheric CO_2 data from the two new sites with the data that we have been using from Mauna Loa. Therefore, we should choose the same resolution data that we used for our Mauna Loa analysis: the monthly resolution data. If I wanted to compare the exact date of the peak values or look at variability within the month, I would choose the higher resolution data. (In that case I would also have to find higher resolution data for Mauna Loa.) But to answer our question about the seasonal cycles and the long-term trend, the monthly resolution is appropriate.

From Barrow, Alaska, the in situ samples are only available from 1961 to 1967, while the flask samples are available from 1974 to the present. For the purposes of this comparison, I am interested in looking at the long-term trend and the seasonal cycle. Therefore, I am going to choose the monthly data for the longest time period: the flask measurements, compiled monthly. For the South Pole, the in situ and flask samples have been merged and are available from 1957 to the present. Again I will choose the compiled monthly data.

As we saw in Chapter 7, for each location there are many different columns of CO_2 values. The metadata describes the columns and indicates that column 5, labelled CO_2 (ppm), is the most commonly sought-after data. However, there are a lot of gaps in the data (evident by the number of place holders: −99.99). This will make a quantitative comparative analysis difficult. The gaps will have to be filled or I will have to choose only a subset of the data that is complete for all months of the year at all three sites over the same timeframe.

Column 7 is described as a "smoothed version of the data generated from a stiff cubic spline function plus four-harmonic functions with linear gain."[2] The original researchers have generated this "smoothed" data set to fill the data gaps. For my current purpose, this data will be easier to work with. I will choose column 7 as the CO_2 data for my comparative analysis.

11.2 COMPARING DATA SETS

The three different CO_2 data sets from different locations can be visually compared using a scatterplot (Figure 11.1). There are noticeable similarities and differences between these three data sets. We can see that an annual cycle and an increasing trend are present in all three data sets. However, the amplitude of the annual cycle expressed in these three data sets is different. If you look closely at Figure 11.1b, you might notice that the annual cycle from the South Pole is antiphase (opposite) of the other two records reflecting the timing of the summer growing season and winter senescence.

The similarities between these data sets, the average concentrations of the three datasets are similar and change over time within the same order of magnitude, lend support to the idea that these data sets are representing a common phenomenon. But the differences, similar but not identical concentrations and antiphase annual cycles, suggest the relationship between these three variables is not straightforward.

We might want to ask, *how much of the change that occurs in one location is mimicked by a similar change in another location*?

This type of analysis, determining how much variability is common between two data sets, is called correlation.

11.2.1 Correlation Analysis

Correlation analysis doesn't tell us anything about whether the change in one data set is responsible or is a cause of the change in another data set (correlation does not equal causation). Mathematically, correlation describes whether there is a similarity in the variability between the

[2] http://scrippsco2.ucsd.edu/data/atmospheric_co2/mlo

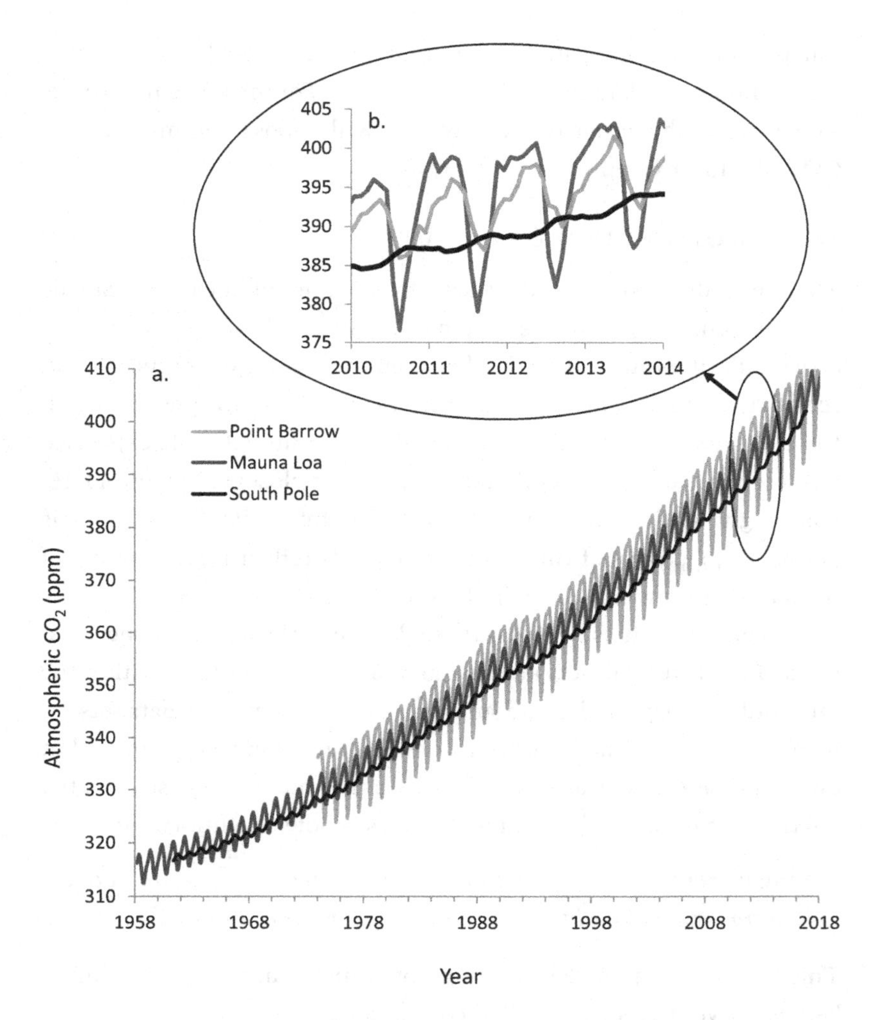

FIGURE 11.1 Atmospheric CO_2 concentrations (ppm) from three different locations: Point Barrow, Alaska; Mauna Loa, Hawaii; South Pole, Antarctica.

Data from Keeling et al. (2001). (a) The full time series from 1958 to 2018. (b) A subset of data from 2010 to 2014.

two data sets of interest: when one data set changes, does the other data set demonstrate a change too? In which direction?

Correlation analysis allows us to determine whether the two data sets change in the same way (a positive relationship), in the opposite way (a negative relationship), or in no related way. The calculated correlation coefficient (r) ranges from 1 to -1:

- A correlation of $r = 1$ indicates that the changes in one data set are perfectly "in phase" with the changes in the other data set (Figure 11.2a). This means that the change in one data set is matched by change in the same direction in the other data set.
- A correlation of $r = -1$ indicates that the changes in one data set are perfectly out of phase with, or opposite to, the other data set (Figure 11.2b).
- A correlation of $r = 0$ indicates that there is no relationship between the two data sets (Figure 11.2c).

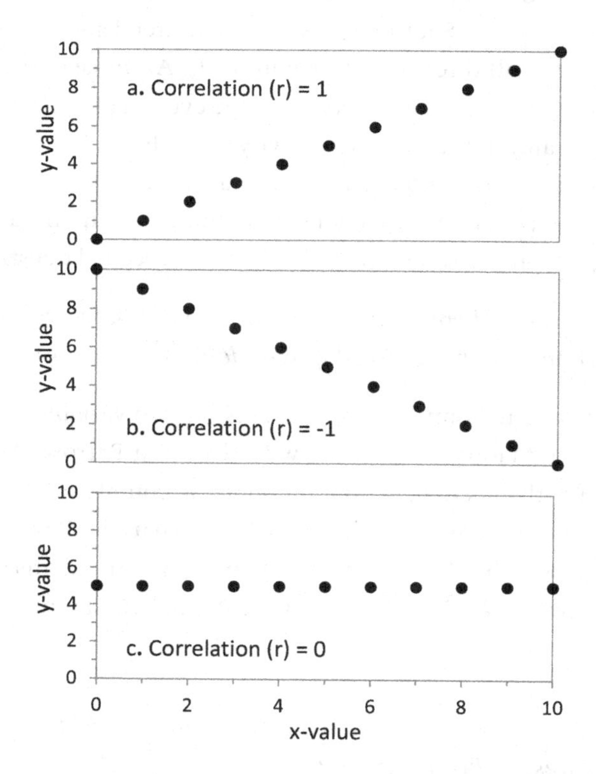

FIGURE 11.2 Schematic representations of different correlation coefficients (r). (a) A theoretical positive correlation. (b) A theoretical negative correlation. (c) A representation of no correlation.

Correlation coefficient values in between 1 and −1 are used to describe the strength of the relationship between the two data sets.

11.2.2 Calculating a Correlation

Correlation is commonly used to compare different variables. However, in this case, we are attempting to answer the question

Are the seasonal cycle and long-term trend, evident in atmospheric CO_2 concentrations collected from Mauna Loa, Hawaii, representative of local or global phenomena?

To investigate this question, I will use the three different records of atmospheric CO_2 that we now have from Mauna Loa (Hawaii), Point Barrow, (Alaska), and the South Pole. A long-term trend and a seasonal cycle are evident in all three records (Figure 11.1). All three of the CO_2 data sets are responding to external forcings: the cycle of photosynthesis and respiration and anthropogenic activity. All three data sets are changing over time. When more than one signal of variability is present, two data sets can contain positively related, unrelated, and negatively related change, all at the same time. This leads to a second question:

In a situation where there is more than one signal being compared at the same time, which relationship will dominate the correlation?

Correlation is used to compare two data sets (or two variables). I will first compare the Mauna Loa data set with the Point Barrow, Alaska, data set. Using the CO_2 data that provides a complete record of atmospheric CO_2 (column 7 in the spreadsheet that we downloaded from the Scripps CO_2 Program in Chapter 7), I can calculate the correlation between the Mauna Loa CO_2 data and the Point Barrow CO_2 data.

11.2.2.1 Step 1: Make Sure That You Are Using the Same Time Frame and Dates for Both Data Sets

The Point Barrow data set spans a shorter amount of time than the Mauna Loa data set, so I have to truncate the Mauna Loa data set to match the dates covered by Point Barrow. For both data sets, the first data point will be January 1968 and the last data point will be December 2017. Before proceeding, I will do a quick check to see that all of the datapoints for the two sites have matching dates.

11.2.2.2 Step 2: Set Up Your Calculation

This is the equation for calculating a Pearson's correlation:

$$r_{xy} = \frac{\Sigma(x_i - x_{avg})(y_i - y_{avg})}{\sqrt{\Sigma(x_i - x_{avg})^2 \Sigma(y_i - y_{avg})^2}}.$$

To make sure that you can easily find errors in your calculations, I encourage you to separate out each step of the calculation into its own column. In Table 11.1, you can see the beginning and the end of my spreadsheet. In the first cell of each calculated column, I have shown you the formula that I used. I then copied the formula into each subsequent cell in the column to execute the calculation. Make sure that you check to see that the formula is formatted properly so that you are referencing the correct cells as you copy a formula down a column.

I have broken down the correlation calculation to do each step of the calculation separately. This makes troubleshooting easier in Excel. Below is a description of the spreadsheet:

- Column A is the decimal year.
- Column B is the Mauna Loa CO_2 monthly data.
- Column C is the difference between the monthly average CO_2 value and the average CO_2 value for the Mauna Loa data set.
  - ⇒ Because the average CO_2 value for Mauna Loa is in cell B606, I used the $ sign to lock both the column and row for this cell.
- Column D is the value from Column C squared.
- Column E is the Point Barrow CO_2 monthly data.
- Column F is the difference between the monthly average CO_2 value and the average CO_2 value for the Point Barrow data set.
  - ⇒ Again, because the average CO_2 value for Point Barrow is in cell E606, I used the $ sign to lock both the column and row for this cell.

Table 11.1 *A spreadsheet demonstrating the correlation coefficient (r) calculations for two CO_2 records: Mauna Loa, Hawaii, and Point Barrow, Alaska*

	A	B	C	D	E	F	G	H	I
1	**Date**	**Mauna Loa CO2**	**CO2-avgCO2**	**(CO2-avgCO2)^2**	**Point Barrow CO2**	**CO2-avgCO2**	**(CO2-avgCO2)^2**		**Mauna Loa (ML), Point Barrow (PtB)**
2	**decimal year**	**monthly avg (ppm)**	**Mauna Loa**	**Mauna Loa**	**monthly avg (ppm)**	**Pt Barrow**	**Pt Barrow**		**(CO2-avgCO2)ML* (CO2-avgCO2)PtB**
3	1968.041	322.61	=B3-$B$606	=C3^2	329.06	=E3-$E$606	=F3^2		=C3*F3
4	1968.126	323.29	−36.26	1,314.48	329.82	−31.57	996.94		1,144.75
5	1968.205	324.09	−35.46	1,257.11	330.05	−31.34	982.47		1,111.34
6	1968.290	325.26	−34.29	1,175.51	330.25	−31.14	969.97		1,067.81
7	1968.372	325.81	−33.74	1,138.10	330.88	−30.51	931.13		1,029.42
8	1968.456	325.17	−34.38	1,181.69	328.99	−32.40	1050.05		1,113.92
9	1968.538	323.79	−35.76	1,278.47	322.96	−38.43	1477.20		1,374.25
10	1968.623	321.91	−37.64	1,416.45	318.29	−43.10	1857.99		1,622.26
11	1968.708	320.38	−39.17	1,533.95	320.12	−41.27	1703.58		1,616.54
12	1968.790	320.42	−39.13	1,530.82	324.62	−36.77	1352.36		1,438.82
	•••	•••	•••	•••	•••	•••	•••		•••
598	2017.622	405.25	45.70	2,088.88	395.61	34.22	1170.71		1,563.80

599	2017.707	403.46	43.91	1,928.47	398.24	36.85	1357.60		1,618.05
600	2017.789	403.51	43.96	1,932.86	404.99	43.60	1900.58		1,916.65
601	2017.874	405.02	45.47	2,067.91	410.13	48.74	2375.16		2,216.22
602									
603									
604		**Avg Mauna Loa CO2**		**Sum (CO2-avgCO2)^2**	**Avg Pt Barrow CO2**		**Sum (CO2-avgCO2)^2**		**Sum ((CO2-avgCO2)ML* (CO2-avgCO2)PtB)**
605		**=average (B3: B601)**		**=sum(D3: D601)**	**=average (E3:E601)**		**=sum(G3: G19)**		**=sum(I3: I601)**
606		359.55		348,749.69	361.39		347,321.69		341,857.92
607									
608				**Square Root (Sum (CO2-avgCO2) ^2)**			**Square Root (Sum (CO2-avgCO2) ^2)**		**Correlation**
609				**=sqrt (D606)**			**=sqrt(G23)**		**=I606/ (D610*G610)**
610				590.55			589.34		**0.98**

Data from Keeling et al. (2001).

- Column G is the value from Column F squared.
- Column I is the difference between the monthly CO_2 values and the average CO_2 value from Mauna Loa (Column C), multiplied by the difference between the monthly CO_2 values and the average CO_2 value from Point Barrow (Column F).

The correlation calculation also employs the square root of the summed values of Column D and G. I calculated the sum and the square root of the sum of the two columns separately in cells D606 and D610 and in cells G606 and G610. Finally, I calculated the sum of Column I in cell I606. The final correlation coefficient (r) is in cell I610.

The correlation between the Mauna Loa monthly CO_2 data and the Point Barrow, Alaska, CO_2 data results in a correlation coefficient (r) of 0.98. When I repeat the correlation calculation using the data from Mauna Loa and the South Pole (adjusted so that the timescales on both data sets match), I get a correlation coefficient (r) of 0.99.

☝ Let's think about this for a minute. We know that the seasonal cycle in the Northern Hemisphere and the Southern Hemisphere are opposite each other, so why is the correlation between Mauna Loa, Hawaii, and the South Pole so strong? This is curious.

11.2.3 Correlating the Seasonal Cycles

To confirm that an anti-phase seasonal cycle exists between the Northern Hemisphere site (Mauna Loa) and the Southern Hemisphere site (South Pole), I will isolate the seasonal cycles from these two locations and repeat the correlation calculation.

As in Chapter 8, I will calculate a running average and remove the long-term trend from each data set by subtracting the running average from the original data set. When the long-term trends are removed, you can see that the residual signals reflecting the seasonal cycles from Mauna Loa and the South Pole are indeed opposite each other (Figure 11.3). The Mauna Loa data set has a peak CO_2 value in May, while the South Pole has a peak CO_2 value in October.

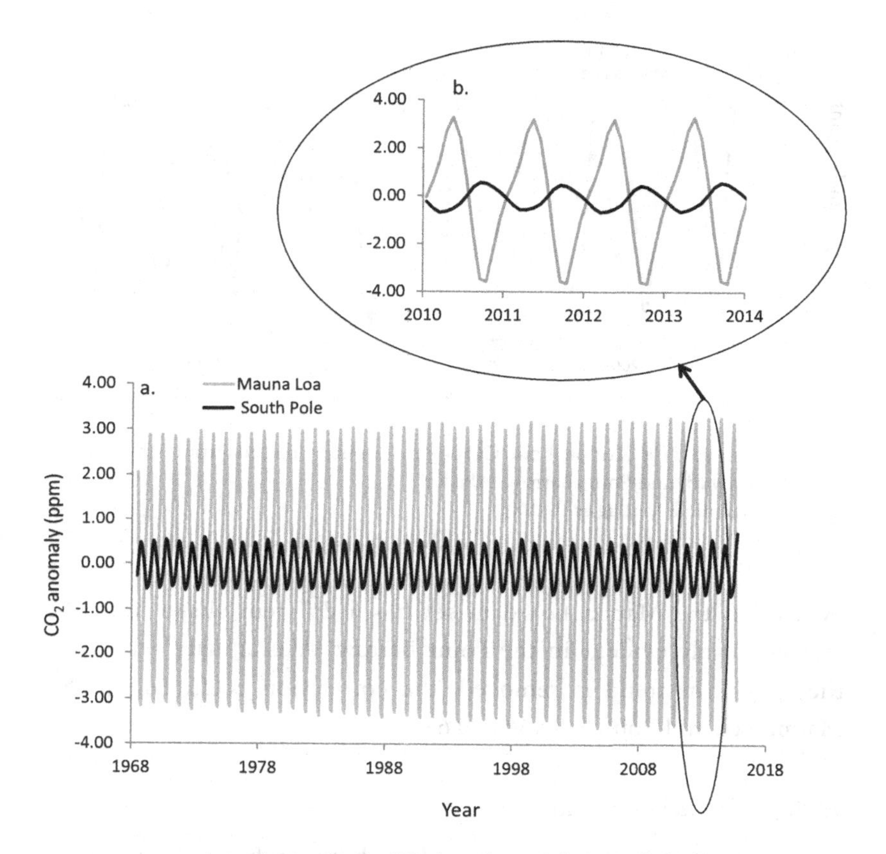

FIGURE 11.3 Atmospheric CO_2 (ppm) anomaly records from Mauna Loa, Hawaii, and the South Pole, Antarctica.
Data from Keeling et al. (2001). (a) The full time series from 1968 to 2017. (b) A subset of data from 2010 to 2014 highlighting the seasonal cycles.

Recalculating the correlation between these two locations, using the isolated seasonal cycle data, results in a correlation coefficient (r) of -0.91. This strong, negative correlation makes sense for the seasonal data: it reflects the anti-phase relationship between these two records that are both changing at the same time (on a seasonal timescale) but in opposite directions.

For a comparison, I also calculated the correlation coefficient (r) for the isolated seasonal cycles from the Mauna Loa data set and the Point Barrow, Alaska, data set (Figure 11.4). The seasonal cycle

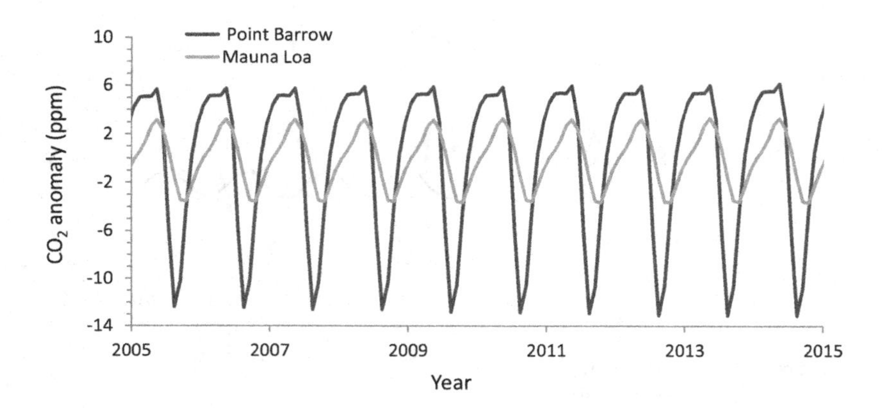

FIGURE 11.4 Atmospheric CO_2 (ppm) anomaly records from Mauna Loa, Hawaii, and Point Barrow, Alaska.
Data from Keeling et al. (2001).

evident in the Mauna Loa data set has a different shape from the seasonal cycle evident in the Point Barrow, Alaska, data set. When the long-term trend is removed, the correlation coefficient (r) for Mauna Loa and Point Barrow is 0.68.

11.2.4 Signal Dominance

The correlation coefficient changed for both analyses, Mauna Loa–South Pole and Mauna Loa–Point Barrow, when I isolated the seasonal signals (Table 11.2).

Why?

In Chapter 8, signal dominance was determined by comparing the amplitudes for each frequency of change present in the data set. In the same way, the amplitude is important when calculating correlations. The correlation between the Mauna Loa and the South Pole monthly average CO_2 values was influenced by the dominant signal: the long-term trend. Therefore, the strong positive correlation coefficient that we calculated reflected the presence of a common dominant signal. The influence of the seasonal cycle was not strong

Table 11.2 *A comparison of correlation coefficients between CO_2 data from Mauna Loa, Hawaii; the South Pole, Antarctica; and Point Barrow, Alaska, calculated on the full data set and on the isolated data (anthropogenic long-term trend removed)*

	Mauna Loa full data set	Mauna Loa seasonal cycle only
Point Barrow full data set	0.98	
Point Barrow seasonal cycle only		0.68
South Pole full data set	0.99	
South Pole seasonal cycle only		−0.91

Data from Keeling et al. (2001)

enough to influence the correlation coefficient calculation in a meaningful way.

However, when the long-term trend was removed from both records, the anti-phase relationship between the Mauna Loa and the South Pole seasonal cycles became evident. The calculated correlation coefficient of −0.91 reflects the strength of the anti-phase relationship in the isolated seasonal signals.

11.3 SUMMARY

In this chapter we investigated the following question:

Are the seasonal cycle and long-term trend, evident in atmospheric CO_2 concentrations collected from Mauna Loa, Hawaii, representative of local or global phenomena?

Using correlation analysis, we compared Mauna Loa atmospheric CO_2 concentrations with atmospheric CO_2 concentrations from Point

Barrow, Alaska, and the South Pole. Using data sets with the same sample resolution and timeframe, strong positive relationships were found between both Mauna Loa and Point Barrow, Alaska, (r = 0.98) and Mauna Loa and the South Pole (r = 0.99). The long-term trend stands out as a common signal among all three data sets, representing a global phenomenon.

When the seasonal cycle is isolated and compared, an anti-phase relationship between Mauna Loa (Northern Hemisphere) and the South Pole (Southern Hemisphere) is evident, with a correlation coefficient (r) of −0.91. The isolated seasonal cycles from Mauna Loa, Hawaii, and Point Barrow, Alaska, (both Northern Hemisphere sites) are in phase but not identical (r = 068). The correlation analysis indicates that the seasonal cycle is location dependent. Intra-hemisphere and inter-hemisphere differences reflect the timing and strength of local photosynthesis and respiration.

11.4 TAKE-HOME MESSAGES

- Correlation analysis examines the strength and direction of the common variability between two data sets.
- Correlation doesn't by itself give information about causation.
- Environmental data sets may contain more than one signal of variability.
- The calculated Pearson's correlation coefficient (r) is influenced by the dominant signal of variability present in the two data sets.

REFERENCE

Keeling, C. D., S. C. Piper, R. B. Bacastow et al. (2001). Exchanges of atmospheric CO_2 and $13CO_2$ with the terrestrial biosphere and oceans from 1978 to 2000. I. Global aspects, SIO Reference Series, No. 01-06, Scripps Institution of Oceanography, San Diego, CA, 88pp.

Chapter 12

Preparation

1. Before reading Chapter 12, consider the following questions
   - How does a system change over time?
   - Can we predict future behaviour based on past data?
   - What is the difference between correlation and regression analysis?
   - How do you start to build a model?
2. Revisit the graphs you drew in Chapter 8. On the graph you drew reflecting the number of students on campus over a year, label:
   a. the periods in time when the number of people coming to campus exceeds the number of people leaving the campus;
   b. the periods in time when the number of people leaving campus exceeds the number of people coming to the campus; and
   c. the periods of time when the number of people coming to campus and the number of people leaving campus are equal.
3. On panel (a), draw a graph that represents the inflow of student over the year. On panel (b), draw a graph that represents the outflow of students over the year.

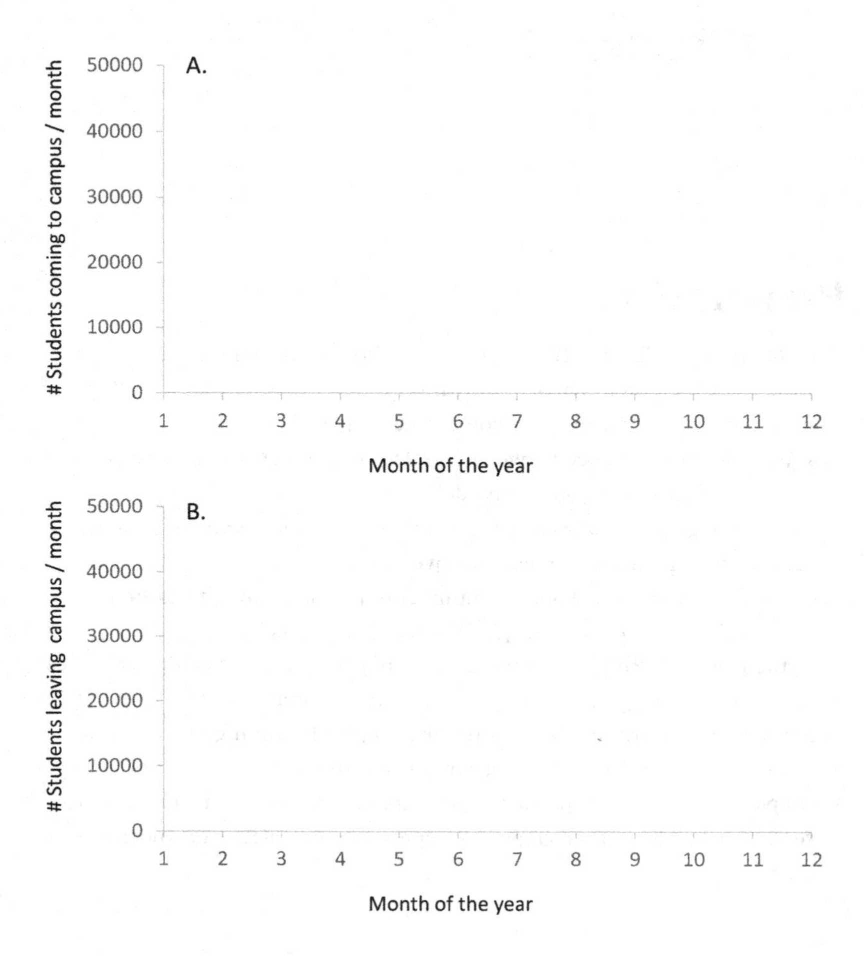
A.
Students coming to campus / month
50000
40000
30000
20000
10000
0
1 2 3 4 5 6 7 8 9 10 11 12
Month of the year
B.
Students leaving campus / month
50000
40000
30000
20000
10000
0
1 2 3 4 5 6 7 8 9 10 11 12
Month of the year

12 Developing Simple Environmental Models

Models are simplified representations of reality. We use models in many different ways, for many different purposes: physical models are used to study urban planning, vehicle impacts, new inventions, and molecular structures; rats are used as model organisms to test new pharmaceuticals for side effects; conceptual models are used to simplify and understand concepts, ideas, and relationships; mathematical models are used to understand, quantify, and predict the behaviour of the natural world. All models are intended to capture some key feature of reality (shape, behaviour, relationships, etc.); the specific focus of a model will depend on the purpose for which the model is intended to be used.

Environmental modelling helps deepen our understanding of how the integrated physical, chemical, and biological systems on Earth have changed over time and how they might change in the future.

12.1 CHANGE OVER TIME

A system that changes over time is called a dynamic system. Dynamic environmental systems can be complex. In Chapter 10 we saw that the seasonal cycle, demonstrated by changes in the record of non-anthropogenic atmospheric CO_2, is not identical every year. The annual variability that is evident in the atmospheric CO_2 data is dominated by plant photosynthesis and respiration. We could speculate on the reasons for variability in annual CO_2 cycle.

Plant photosynthesis could be influenced by

- the number of plants present,
- the type of plants present,

- the amount of water and sunlight that each plant receives, and
- the amount of nutrients (macro and micro) available to the plants during the growing season.

Plant respiration could be influenced by

- the amount of water, oxygen, and CO_2 available;
- the temperature the plants experience; and
- the life stage of the plant.

Each of these factors will, in turn, be influenced by other aspects of the environment. For example, at many locations, temperature, cloud cover, and precipitation are influenced by systems like the El Niño / Southern Oscillation or regional monsoons.

At any one time, in any one location, the value measured for atmospheric CO_2 is determined by all of the individual contributions to the collective regional ecology.

In Chapter 10 we characterized the inter-annual variability of atmospheric CO_2 for each month using the anomaly data set that was generated by subtracting the long-term trend from the full data set. The standard deviations of the Mauna Loa monthly CO_2 anomalies from 1960 to 2017 were calculated (Figure 10.11). Though peak CO_2 values occur in May every year, the May values are not the same every year. Plotting only the May CO_2 anomaly values between 1959 and 2017, you can see the inter-annual variability (Figure 12.1). You might also notice that over the entire timescale we are considering, the average May values have drifted higher. Not every May value is higher than the previous value, but overall the May values are progressively getting higher over time.

May is the transition month between the Northern Hemisphere dormant period and the onset of the growing season. The increasing May values over time imply that the amount of CO_2 released to the atmosphere from respiration during winter is increasing.

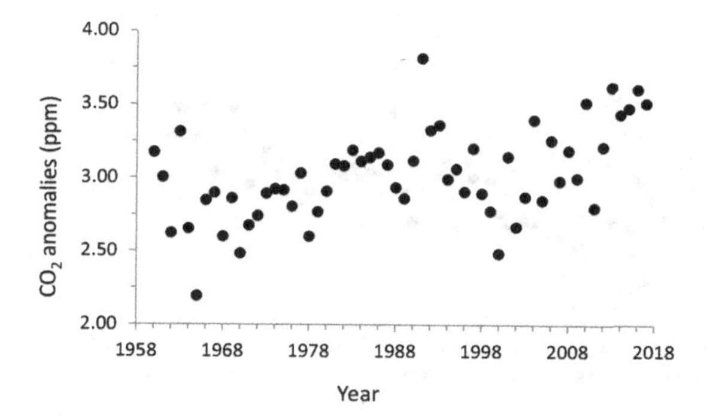

FIGURE 12.1 May CO_2 anomaly values between 1959 and 2017. Data from Keeling et al. (2001).

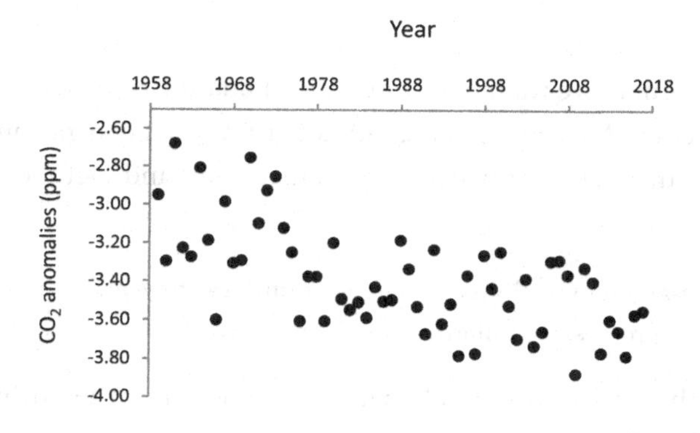

FIGURE 12.2 October CO_2 anomaly values between 1959 and 2017. Data from Keeling et al. (2001).

In contrast, the average October CO_2 concentrations at Mauna Loa are decreasing over time (Figure 12.2). This suggests that at Mauna Loa, the amount of CO_2 being removed from the atmosphere during the summer growing season is increasing. Calculating the annual difference between the average May CO_2 values and the average October CO_2 values (Figure 12.3) is another way of investigating changes in the amplitude of the seasonal atmospheric CO_2 cycle.

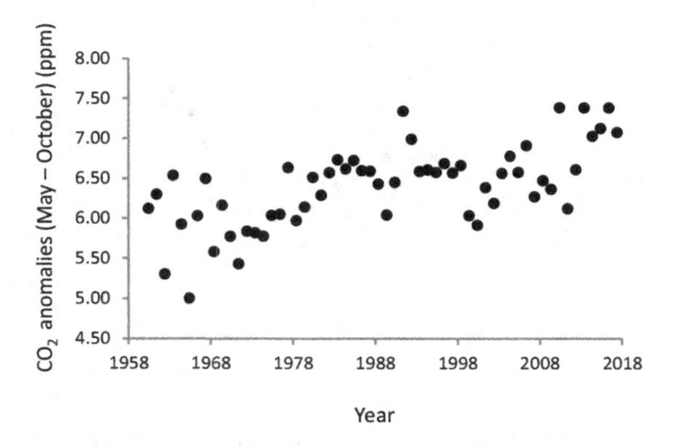

FIGURE 12.3 May–October CO_2 anomaly values between 1959 and 2017. Data from Keeling et al. (2001).

Once again, more than one timescale of change is present in the records of average May and average October CO_2 concentrations, as well as in the difference between the average May and October CO_2 concentrations:

- There is inter-annual variability – a change from year to year.
- There is also a progressive unidirectional change over time.

In this case, the unidirectional change over time is not the anthropogenic signal that we saw before. The anthropogenic signal was removed when we created the anomaly data set to isolate the annual cycle. The unidirectional change over time that is present here tells us something about plant ecology, not anthropogenic activity.

12.2 MODELLING A TREND

Mathematical models represent relationships in a system in the form of equations that relate variables. We can use mathematical models to investigate how changing one (or more) variable(s) changes the system as a whole over time.

Can we use a simple model to represent the trends that are present in our data set?

Occam's razor, a foundational philosophical concept in science, celebrates simplicity by suggesting that the best solution is the one that relies on the smallest number of assumptions. A linear model is the simplest model that can be used to represent a system. Depending on your question, a linear model might provide a good answer. In this case, because we can see a consistent unidirectional change increasing over time, we can consider using a linear model to investigate our question. We might ask:

Does a linear model explain the increasing trend seen in the May atmospheric CO_2 anomalies from 1959 to 2017?

12.2.1 Calculating a Line of Best Fit

Using the least squares methods (see Chapter 7), the equation for the line that minimizes the difference between my measured y-values (May anomalies) and the modelled y'-values (that are predicted by a line of best fit) is

$$y' = 0.01x - 17$$

To find the slope (m) and intercept (b) of the line of best fit, I made a spreadsheet in Excel that isolates each critical step into separate columns (Table 12.1). I repeated this process for the October CO_2 anomaly data ($y' = -0.001x + 16$) and the May–October difference anomalies ($y' = 0.02x - 33$) for each year (Figure 12.4).

12.2.2 Intuitively Evaluating a Linear Model

How much of the variability present on the data do these linear models explain?

The first step in determining if a linear model makes sense for your data is to check to see

Table 12.1 *A spreadsheet demonstrating the line of best fit (linear regression) calculations for the May CO_2 anomalies from 1960 to 2017*

	A	B	C	D	E	F	G	H	I	J
1	**Avg X =**	**=AVERAGE (D2:D60)**		**X = Date**	**Y = May anomalies**	**X-AvgX**	**(X-AvgX) ^2**	**Y-avgY**	**(X-AvgX)* (Y-avgY)**	**Regression line**
2	**Avg X =**	**1988.37**		1959.3699	2.38	**=D2-B2**	**=F2^2**	**=E2-B5**	**=F2*H2**	**=$B$8*D2 +$B$11**
3				1960.3716	3.18	−28.00	783.93	0.17	−4.89	2.72
4	**Avg Y =**	**=AVERAGE (E2:E60)**		1961.3699	3.00	−27.00	729.02	0.00	0.08	2.73
5	**Avg Y =**	**3.01**		1962.3699	2.62	−26.00	676.02	−0.38	9.96	2.74
6				1963.3699	3.31	−25.00	625.02	0.31	−7.63	2.75
7	**b = slope =**	**=I63/G63**		1964.3716	2.66	−24.00	575.94	−0.35	8.30	2.76
8	**b = slope =**	**0.01**		1965.3699	2.19	−23.00	529.02	−0.81	18.72	2.77
9				1966.3699	2.84	−22.00	484.02	−0.16	3.57	2.78
10	**a = intercept =**	**=C5-(C8*C2)**		1967.3699	2.90	−21.00	441.02	−0.11	2.30	2.79
11	**a = intercept =**	**−17.07**		1968.3716	2.60	−20.00	399.95	−0.41	8.16	2.80

12				1969.3699	2.86	−19.00	361.02	−0.15	2.76	2.81
	...	...		...	...	...	...	...	...	...
50				2007.3699	2.98	19.00	360.98	−0.03	−0.50	3.20
51				2008.3716	3.19	20.00	400.05	0.18	3.59	3.21
52				2009.3699	3.00	21.00	440.98	−0.01	−0.13	3.22
53				2010.3699	3.51	22.00	483.98	0.51	11.15	3.23
54				2011.3699	2.80	23.00	528.98	−0.21	−4.82	3.24
55				2012.3716	3.21	24.00	576.06	0.20	4.91	3.25
56				2013.3699	3.62	25.00	624.98	0.61	15.34	3.26
57				2014.3699	3.43	26.00	675.98	0.42	11.04	3.27
58				2015.3699	3.47	27.00	728.98	0.47	12.61	3.28
59				2016.3716	3.60	28.00	784.07	0.60	16.74	3.29
60				2017.3699	3.50	29.00	840.97	0.50	14.44	3.30
61										
62						**Sum =**	**=Sum (G2:G60)**	**Sum=**	**=Sum (I2:I60)**	
63						**Sum =**	**17,110.00**	**Sum=**	**172.74**	

Data from Keeling et al. (2001).

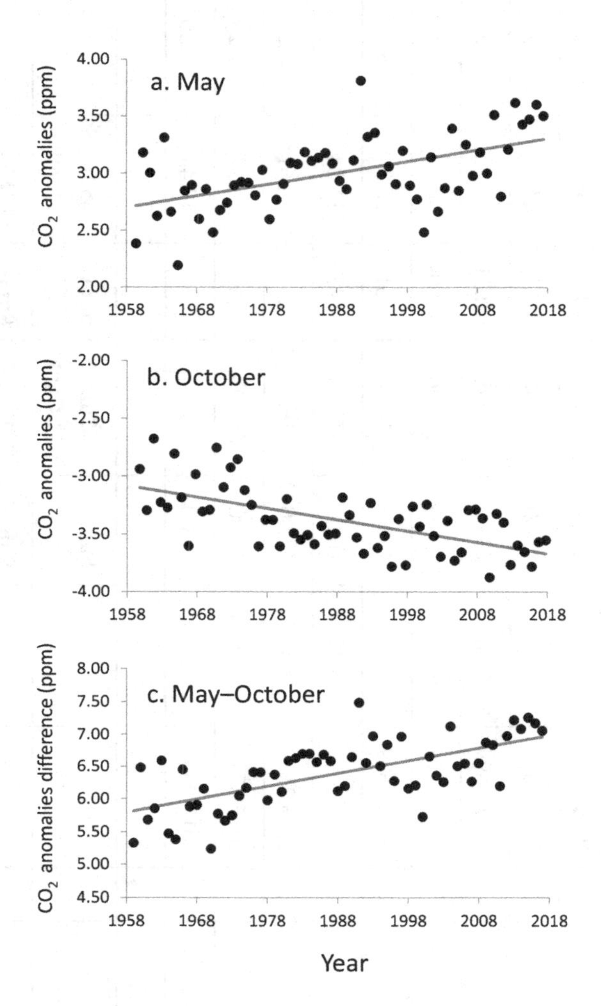

FIGURE 12.4 Linear regression models of the (a) May, (b) October, and (c) May–October CO_2 anomaly values between 1959 and 2017.

1. that the data is randomly distributed around the line of best fit: approximately half of the data above the line and half of the data below, and
2. each datapoint is independent: you should not see a pattern emerging in the distribution of the datapoints around the line.

If you see that there is an obviously non-linear aspect to your data, a linear fit might not be the best model to choose. In this case, no obvious patterns are evident when the May, October, and May–October difference CO_2 data from Mauna Loa are plotted with linear regressions (Figure 12.4) so a linear fit appears to make sense.

The linear regression models for the May, October, and May–October difference in non-anthropogenic CO_2 concentrations seem to capture the shape of the long-term trend reasonably well. *Can these linear models be used to predict non-anthropogenic CO_2 in future (or past) Mays or future (or past) Octobers?*

I can check this intuitively by looking at the amplitude of the high-frequency variability in the data compared to the change associated with the trend (Figure 12.5). As in Chapter 9, if the amplitude of the high-frequency variability is larger than the change associated with the linear trend over the timeframe in question, you can intuitively understand that the trend line will not be able to explain all (or even most) of the variability in the data set. If the amplitude of the high-frequency variability is small compared to the change associated with the linear trend, then you can be confident that your trend line is capturing the essential feature of the data set.

In all three of our examples, the amplitude of the high-frequency variability is greater than the change associated with the linear regression over the timeframe that we are looking at (1958–2018) (Figure 12.5). We can see the variability in the system is not all explained by a linear model. How much is?

- May anomalies: The change associated with the linear regression is about half that of the amplitude of the high-frequency variability seen in the May anomalies, so the linear regression appears to explain only about one-third of the total variability present in this data set.
- October anomalies: The change associated with the linear regression is just over half of the amplitude of the high frequency variability seen in the October anomalies, so the linear regression appears to explain just over one-third of the total variability present in this data set.
- May–October anomalies: The change associated with the linear regression is nearly equal to the amplitude of the high-frequency variability seen in the

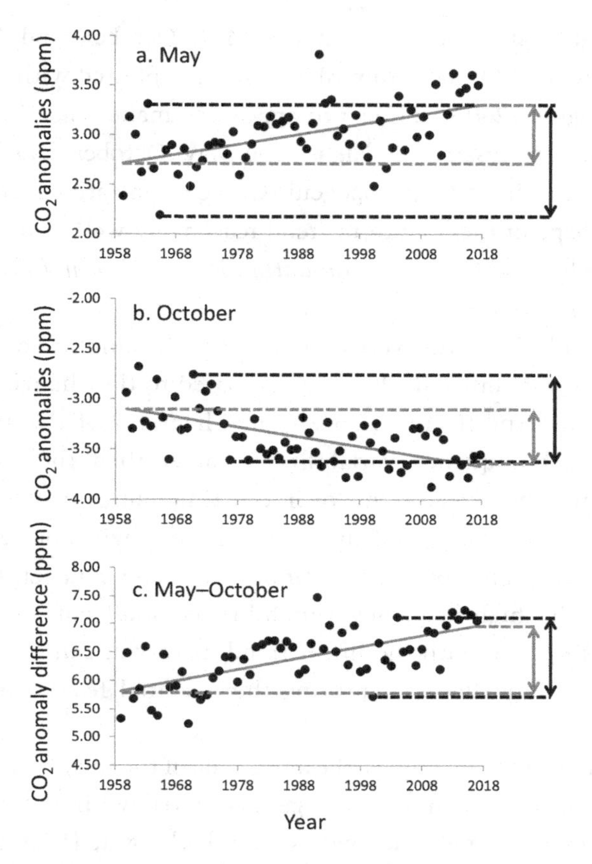

FIGURE 12.5 Estimating the variability explained by the linear regression models for (a) May, (b) October, and (c) May–October CO_2 anomaly values. The black dashed lines approximate the high-frequency variability. The grey dashed lines approximate the change associated with the linear trend from 1958 to 2018.

May–October anomalies, so the linear regression appears to explain nearly half of the total variability present in this data set.

12.2.3 Calculating a Linear Regression Coefficient

To statistically quantify how much of the dependent variable variability that the independent variable is able to predict in each case, we can calculate the regression coefficient of determination (r^2). The coefficient of determination (r^2) is a measure of how much of the variability in the data is "explained," but not "caused," by the linear regression. The coefficient of determination (r^2) is calculated as

$$r^2 = \frac{\Sigma(y_i - y_{avg})^2}{\Sigma(y_{ireg} - y_{avg})^2}$$

- $\Sigma(y_i - y_{avg})^2$ = the total variability in the data set
- $\Sigma(y_{ireg} - y_{avg})^2$ = the variability related to the regression line

I have demonstrated the calculation in Table 12.2 and summarized the results for May, October, and May–October in Table 12.3.

How should you interpret r^2, the coefficient of determination?

In our May example, by comparing the amplitude of the variability associated with the long-term trend with the amplitude of variability associated with the high-frequency signal, we intuitively guessed that the linear regression would explain approximately one-third of the variability in the data set. The regression coefficient of determination (r^2) is a way to calculate the amount of variability explained by the linear regression. The coefficient of determination (r^2) = 0.29 for the May anomaly data set. This means that the linear regression explains 29% of the variability in the data. That is pretty close to our guess.

The calculated r^2 values for the three linear models confirm the visual evaluation of the data:

- Guess: the May linear regression explains about one-third of the total variability present in this data set.
  ✓ May r^2 = 0.29; the linear regression explains 29% of the variability.
- Guess: October linear regression explains just over one-third of the total variability present in this data set.
  ✓ October r^2 = 0.38; the linear regression explains 38% of the variability.
- Guess: the May–October linear regression explains nearly half of the total variability present in this data set.
  ✓ May–October r^2 = 0.44; the linear regression explains 44% of the variability.

Table 12.2 *A spreadsheet demonstrating the regression coefficient (r^2) calculations for the May CO_2 anomalies from 1960 to 2017*

	A	B	C	D	E	F	G
1	**X (year)**	**Y (May anomalies)**	**Y′ regression**	**Y-avgY**	**(Y-avgY)^2**	**Y′-avgY**	**(Y′-avgY)^2**
2	1959.3699	2.38	2.71	−0.62	0.39	−0.29	0.09
3	1960.3716	3.18	2.72	0.17	0.03	−0.28	0.08
4	1961.3699	3.00	2.73	0.00	0.00	−0.27	0.07
5	1962.3699	2.62	2.74	−0.38	0.15	−0.26	0.07
	•••	•••	•••	•••	•••	•••	•••
59	2016.3716	3.60	3.29	0.60	0.36	0.28	0.08
60	2017.3699	3.50	3.30	0.50	0.25	0.29	0.09
61							
62	Average y =	=AVERAGE(B2:B60)		sum=	=SUM(E2:E60)	sum=	=SUM(G2:G60)
63	Average y =	3.01		sum=	5.98	sum=	1.74
64							
65	Regression (r^2) =	=G63/E63					
66	Regression (r^2) =	0.29					

Data from Keeling et al. (2001).

Table 12.3 *A comparison of the correlation coefficients* (r) *and the regression coefficients* (r^2) *and for the May, October, and May–October linear regression models*

X-variable	*Y*-variable	Correlation (*r*)	Regression (*r*^2)
Year	May anomalies	0.54	0.29
Year	October anomalies	−0.62	0.38
Year	(May–Oct) anomalies	0.67	0.45

It is evident that the phenomena that we are looking at here – changes in annual photosynthesis and respiration processes – are complex. The regression models are reflecting the large variance that is associated with this complex system. Only a portion of the variability in each case is explained using a linear model.

12.2.4 The Difference between Correlation Coefficients (r) and Regression Coefficient of Determination (r^2)

In Chapter 11 we considered the correlation between the CO_2 concentrations measured at three different locations. Table 12.3 summarizes the correlation and regression coefficients for the May, October, and May–October CO_2 anomaly data sets. It is important to recognize the difference in meaning between the correlation coefficient (r) and the linear regression coefficient (r^2):

- Correlation analysis describes the relationship between the two variables; the correlation coefficient (r) is the measure of the strength and direction of the relationship.
- Regression analysis generates a model that explains the variability of the data set that is associated with the long-term trend. The regression coefficient of determination (r^2) describes the amount of variability explained by the regression model. The regression coefficient of determination (r^2) can be used to determine the predictive capabilities of the model.

Our calculated correlation coefficient for the May data set (r = 0.54) indicates that there is an identifiable positive relationship between the May non-anthropogenic CO_2 values and time. This means that increases in time are generally associated with increases in the May non-anthropogenic CO_2 values. The regression coefficient (r^2 = 0.29) indicates that the linear regression model explains 29% of the data variability.

Our calculated correlation coefficient for the October data set (r = −0.61) indicates that there is an identifiable negative relationship between the October non-anthropogenic CO_2 values and time. This means that increases in time are generally associated with decreases in the October non-anthropogenic CO_2 values. The regression coefficient, r^2 = 0.38, indicates that 38% of the data set variability is explained by the linear regression model.

Finally, the calculated correlation coefficient for the May–October data set (r = 0.67) indicates that there is an identifiable positive relationship between the October non-anthropogenic CO_2 values and time. The regression coefficient, r^2 = 0.45, indicates that 45% of the data set variability is explained by the linear regression model.

12.3 USING ONE VARIABLE TO PREDICT ANOTHER

In the example above, we were considering if the May, October, and May–October anomaly linear regressions could confidently be used to extrapolate future or past data. Commonly, linear regressions are used to predict one variable using another. As we have been working with data that represents the anthropogenic component of atmospheric CO_2, we could ask:

Can annual global anthropogenic CO_2 emissions be used to predict the inter-annual increases in atmospheric CO_2 that are associated with the isolated long-term trend?

To answer this question, we need to

1. Find annual global anthropogenic emissions data
   ✓ Global CO_2 emissions data from fossil-fuel burning, cement manufacturing, and gas flaring is provided by the US Department of Energy, Carbon Dioxide Information Analysis Centre (CDIAC).[1] The annual global emissions of CO_2 are plotted in Figure 12.6.
2. Match the resolution of the atmospheric CO_2 data to the emissions data
   ✓ The CO_2 emissions data is annual, so annual averages of atmospheric CO_2 must be calculated from the monthly average CO_2 data. The global anthropogenic CO_2 emissions data that I have starts in 1751 and ends in 2014. The first full year of the Mauna Loa monthly average CO_2 concentrations from which I can calculate an annual average is 1959. I will match the time frames of both data sets by starting in 1959 and ending in 2014.
3. Calculate the inter-annual change of CO_2
   ✓ As the global anthropogenic CO_2 emission data is "per year," it is best to compare it to the atmospheric change in CO_2 per year. The inter-annual change in atmospheric CO_2 can be calculated by subtracting atmospheric CO_2 concentration (ppm) in one year from the atmospheric CO_2 concentration (ppm) of the next year. The inter-annual change in atmospheric CO_2 concentrations, measured at Mauna Loa, was plotted in Figure 9.6 and is reshown here for convenience as Figure 12.7.
4. Model the linear relationship between annual global anthropogenic CO_2 emissions (the independent variable) and the inter-annual change in atmospheric CO_2 (the dependent variable).
   ✓ The linear regression (Figure 12.8), calculated using global anthropogenic CO_2 emissions per year as the x-value and annual change in atmospheric CO_2 as the y-value is, $y' = 0.06x + 0.2$.
5. Calculate the regression coefficient for the linear regression model
   ✓ $r^2 = 0.5$

The simple linear regression model, using annual CO_2 emissions (fossil-fuel burning, cement manufacturing, and gas flaring) to predict

[1] http://cdiac.ess-dive.lbl.gov/trends/emis/meth_reg.html

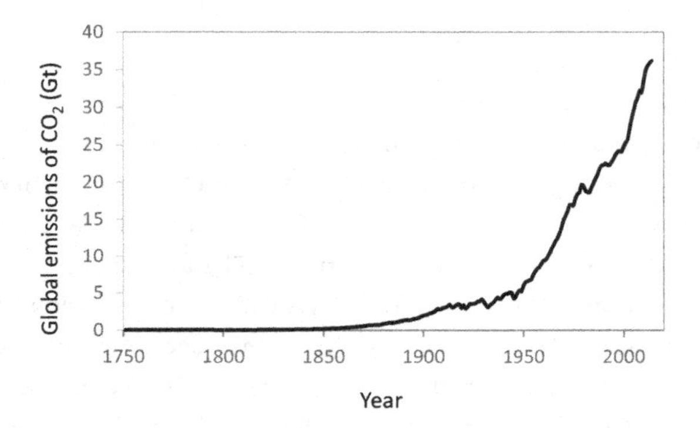

FIGURE 12.6 Global CO_2 emissions (Gt) per year from fossil-fuel burning, cement manufacturing, and gas flaring.
Data is from the US Department of Energy, Carbon Dioxide Information Analysis Centre. (2018). http://cdiac.ess-dive.lbl.gov/trends/emis/meth_reg.html

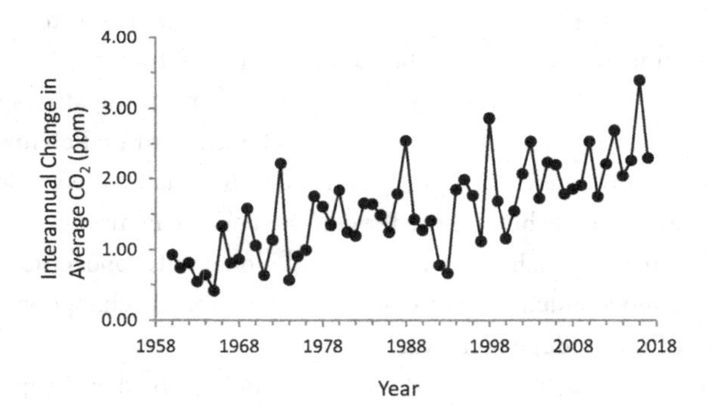

FIGURE 12.7 The inter-annual change in atmospheric CO_2 from 1958 to 2018 measured at Mauna Loa (originally shown as Figure 9.6).
Data from Keeling et al. (2001). http://scrippsco2.ucsd.edu/data/atmospheric_co2/mlo.

inter-annual change in atmospheric CO_2 concentrations, explains 50% of the inter-annual atmospheric CO_2 variability from 1958 to 2017.

What is influencing the other 50% of the variability?

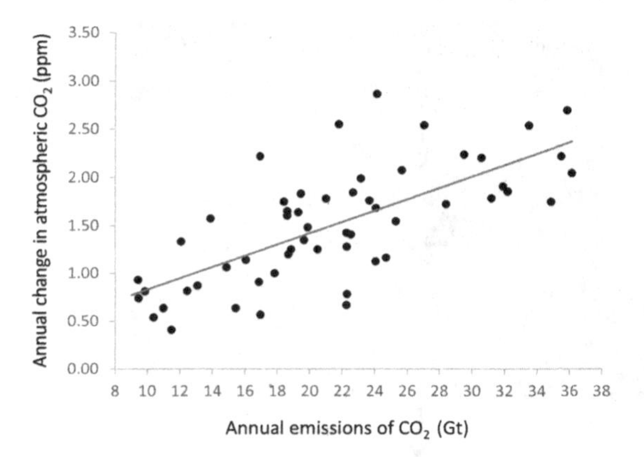

FIGURE 12.8 A linear regression model relating annual anthropogenic emissions of CO_2 (from fossil-fuel burning, cement manufacturing, and gas flaring) to annual increases in the concentration of atmospheric CO_2 (ppm).

Emissions data is from the US Department of Energy, Carbon Dioxide Information Analysis Centre. (2018). http://cdiac.ess-dive.lbl.gov/trends/emis/meth_reg.html. Mauna Loa CO2 concentration data is from http://scrippsco2.ucsd.edu/data/atmospheric_co2/mlo.

12.4 BOX MODELS: STOCK AND FLOW

The linear regression model, describing the inter-annual change in atmospheric CO_2 concentrations from 1958 to 2017, is a simplified representation relating two variables: annual anthropogenic CO_2 emissions and annual changes in atmospheric CO_2 concentrations. This simplification neglects to consider other processes that might influence the average annual concentration of atmospheric CO_2.

Stock and flow models allow multiple processes to be related. A stock is the amount of *something* in a *defined space* over a *defined time*. The concentration of CO_2 in the atmosphere every year is an example of a stock. A flow is a process that causes a stock to either grow or decline over time.

- An *inflow* is a process that adds to the stock. The annual emissions of CO_2 are an example of an inflow.
- An *outflow* is a process that removes stock. A stock can increase or decrease over time if the flows change.

FIGURE 12.9 Stock and flow modelling. Example stock: the amount of water in a sink. Example inflow: the water entering a sink through the tap. Example outflow: water leaving a sink through the drain.

The inflow and outflow processes are like two dials that can be turned up or down such that the stock can increase (or decrease) in a variety of ways.

Think of a sink. You can manipulate the water level in the sink (i.e., the stock) by changing the inflow through the tap or the outflow through the drain (Figure 12.9). The water level in the sink (stock) will increase if

- the inflow *increases* and the outflow *remains the same*,
- the inflow *remains the same* but the outflow *decreases*, or
- both the inflow and outflow increase but the inflow increases proportionally more.

Similarly, the water level in the sink (stock) can be decreased in a variety of different ways. The water level in the tub will decrease if

- the outflow *increases* and the inflow *remains the same*,
- the outflow *remains the same* but the inflow *decreases*, or
- both the outflow and inflow decrease but the outflow decreases proportionally more.

If the inflow is equal to the outflow, then the water level in the sink will remain constant even though there is water running into the sink and water running out of the sink. The level of the water (i.e., the stock) can be in equilibrium at many different heights. The level of water in the sink is dependent on the amount of water entering the basin (inflow) and the amount of water exiting the basin (outflow). Whenever the sum of the inflows balances the sum of the outflows, there will be no change in the stock:

- change in stock = inflow + outflow
- if inflow = outflow, change in stock = 0

We can relate the stocks and the flows of our system using a convention of boxes (for the stocks) and arrows (for the flows). This convention, the conceptual model, communicates clearly what is in the model (and by absence, what is not in the model). The conceptual model also indicates how the stocks and flows are related (Figure 12.10). Building a conceptual model is the first step to building a mathematical model. No model can include everything, so critical decision-making happens at the conceptual model stage. Conceptual models present the thinking that underlies mathematical models. As models become more complex, the conceptual model becomes even more important as a communication tool.

12.4.1 A Stock and Flow Model for Atmospheric CO_2

Consider the amount of non-anthropogenic CO_2 in the atmosphere every month as a stock (Figure 12.11). In this case the inflow would be respiration, a biological process that adds CO_2 to the atmosphere.

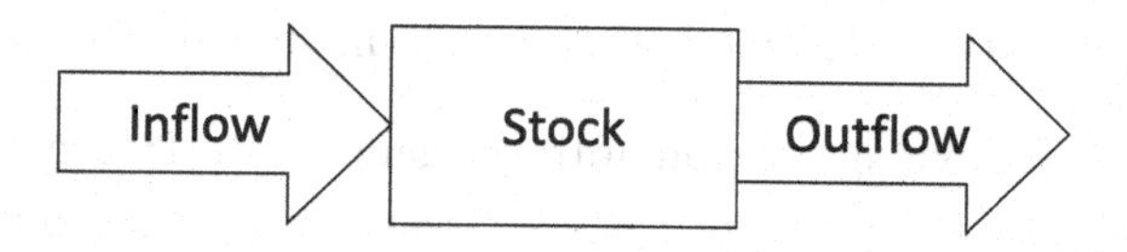

FIGURE 12.10 Stock and flow conceptual model: stocks are represented by boxes; flows are represented by arrows.

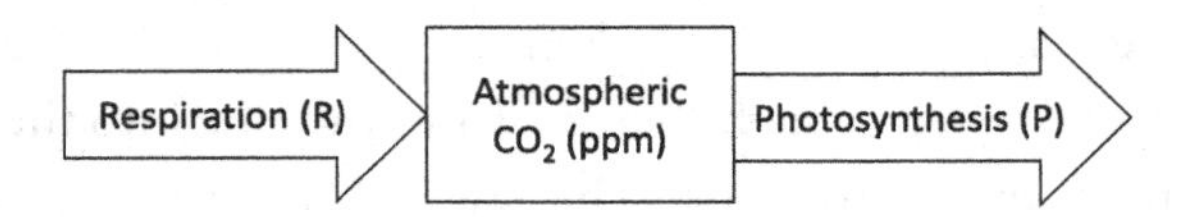

FIGURE 12.11 Conceptual Model: Stock = monthly average atmospheric CO_2 concentrations. Inflow = respiration (R). Outflow = photosynthesis (P).

The outflow would be photosynthesis, a biological process that removes CO_2 from the atmosphere.

Looking at the Mauna Loa average monthly CO_2 data from 1958 to 2017, we can see that from month to month the stock of non-anthropogenic CO_2 in the atmosphere increases and then decreases over one year (Figure 12.12a). Over the summer months, the stock declines, so during the summer the outflow (photosynthesis) must be greater than the inflow (respiration). During the winter months, the stock increases, so during the winter the inflow (respiration) must be greater than the outflow (photosynthesis). The maximum and minimum values of the stock occur in May and October (Figure 12.12b). During these transitional periods, these instances where the change in the stock is 0, the inflow (respiration) and the outflow (photosynthesis) must be equal.

Recall that a stock is the amount of *something* in a *defined place* in a *defined time*. When the stock is defined as the monthly concentration of non-anthropogenic CO_2, there are 12 datapoints a year representing the average concentration of non-anthropogenic CO_2 in the atmosphere (over Mauna Loa) every month. Over this timeframe, seasonal variability is evident.

What if the timescale of the stock was one year instead of one month?

Plotting the yearly average of non-anthropogenic CO_2 (Figure 12.13) shows that over one year the inflow (respiration) and outflow (photosynthesis) are basically balanced, resulting in a stock that is not really changing very much over the 60-year timeframe of our data:

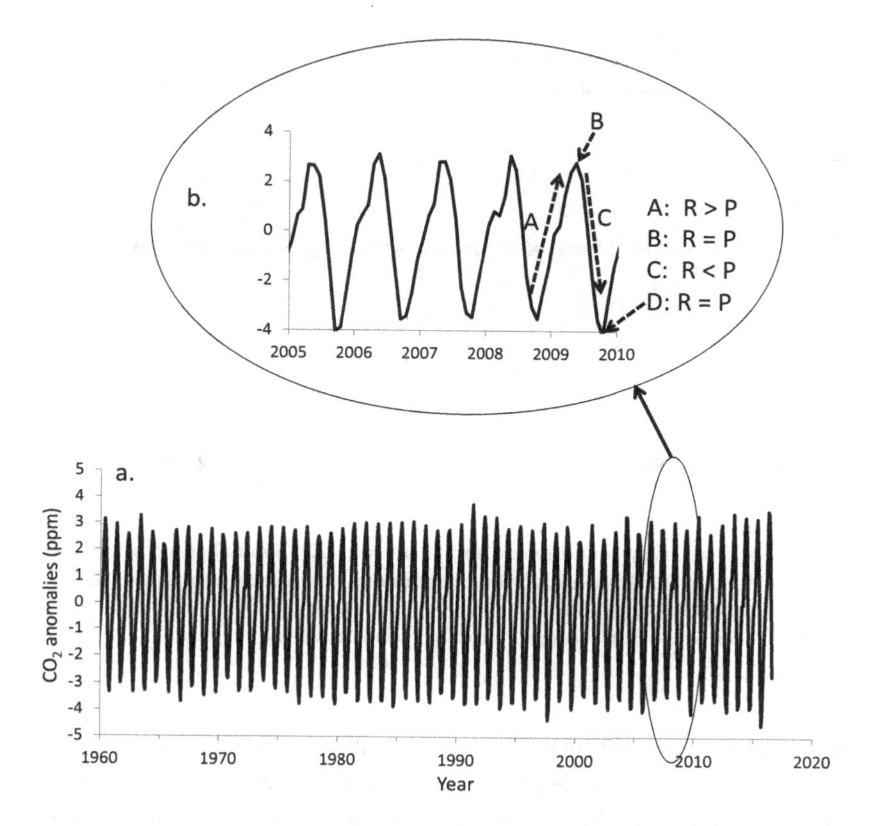

FIGURE 12.12 Monthly average CO_2 anomalies (ppm) from 1960 to 2018. (a) Monthly average CO_2 anomalies (ppm) were generated by removing a 12-month running mean. (b) Monthly average CO_2 anomalies (ppm) from 2005 to 2010. P = photosynthesis, R = respiration.
Data is from http://scrippsco2.ucsd.edu/data/atmospheric_co2/mlo.

$$\text{Annual change in non-anthropogenic } CO_2 = \text{Inflow} + \text{outflow} = 0$$

Matching timescales and processes is a critical aspect of environmental data analysis and modelling. Figure 12.11 demonstrates that processes like photosynthesis and respiration influence seasonal differences in atmospheric CO_2. But seasonal photosynthesis and respiration do not really change atmospheric CO_2 on a decadal or multi-decadal timescale. Unless some significant change in the total

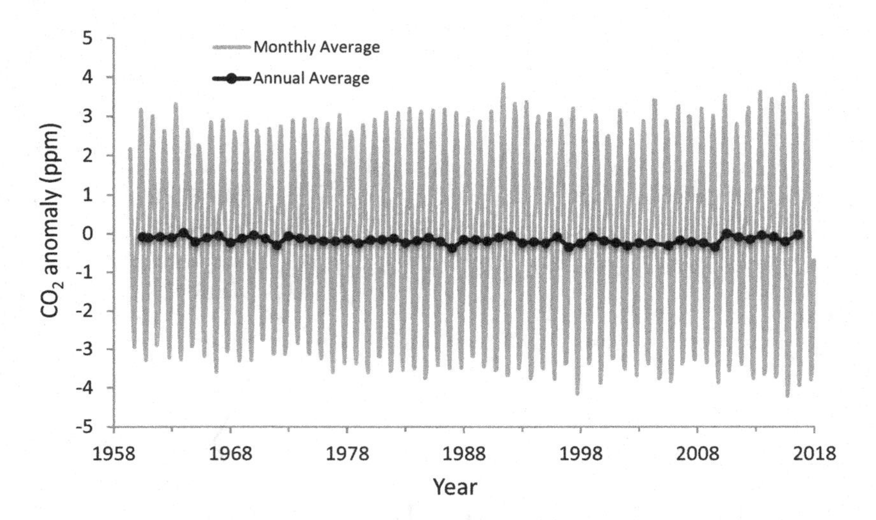

FIGURE 12.13 Monthly and annual average CO_2 anomalies (ppm) from 1960 to 2018. Monthly average anomalies are in grey; annual average anomalies are in black.
Data is from http://scrippsco2.ucsd.edu/data/atmospheric_co2/mlo.

global amount of photosynthesis and respiration were to occur (perhaps through massive deforestation or during a glaciation), the processes of photosynthesis and respiration are most evident on a seasonal timescale.

Considering only the non-anthropogenic CO_2 gives us insight into the seasonal and the inter-annual variability of the biologically driven inflow and outflow of CO_2 to the atmosphere. But we know that

1. natural biological processes are not the only input to atmospheric CO_2, and
2. the complete CO_2 data set from Mauna Loa shows a consistently increasing long-term trend.

So this simple conceptual stock and flow model is fine if we are focused solely on changes in photosynthesis and respiration, but it doesn't capture a major aspect of present day atmospheric CO_2 concentrations – the increasing trend over time. For the conceptual model to reflect the changes in the total stock of atmospheric CO_2, more inflows (sources) and outflows (sinks) of atmospheric CO_2 (and maybe even add more stocks) need to be added.

12.4.2 Adding More Boxes

A first step to more accurately modelling the original Mauna Loa atmospheric CO_2 data set could be to add anthropogenic sources of atmospheric CO_2 into our conceptual model. The revised conceptual model (Figure 12.14) shows that CO_2 inflows to the atmosphere from the biosphere and from burning fossil fuels can be considered separately (as we have been doing in Chapters 7–10). The revised conceptual model also indicates that the concentration of CO_2 in the atmosphere is influenced by both of these processes acting separately but simultaneously.

The level of complexity in a model should be directed by the complexity of the questions you are trying to answer. The Intergovernmental Panel on Climate Change (IPCC), an international body of scientists and government representatives, summarizes our collective understanding of climate change and carbon cycling every five years. In 2014 the IPCC published a model of atmospheric CO_2 including marine uptake and degassing of carbon from the ocean, land use changes, rock weathering, and volcanism (Figure 12.15). With all

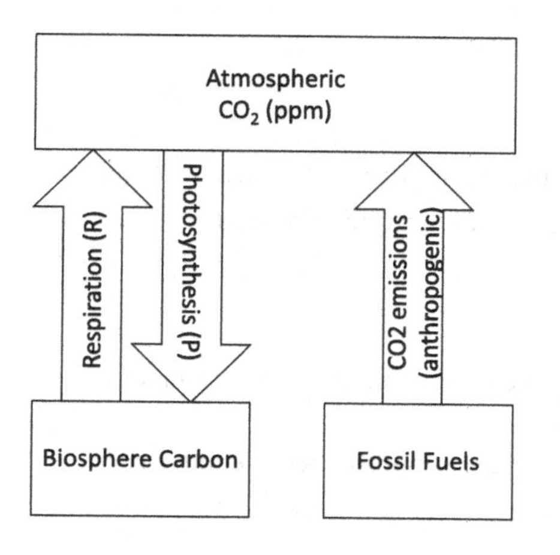

FIGURE 12.14 Influences on atmospheric carbon concentrations from the biosphere and anthropogenic activity.

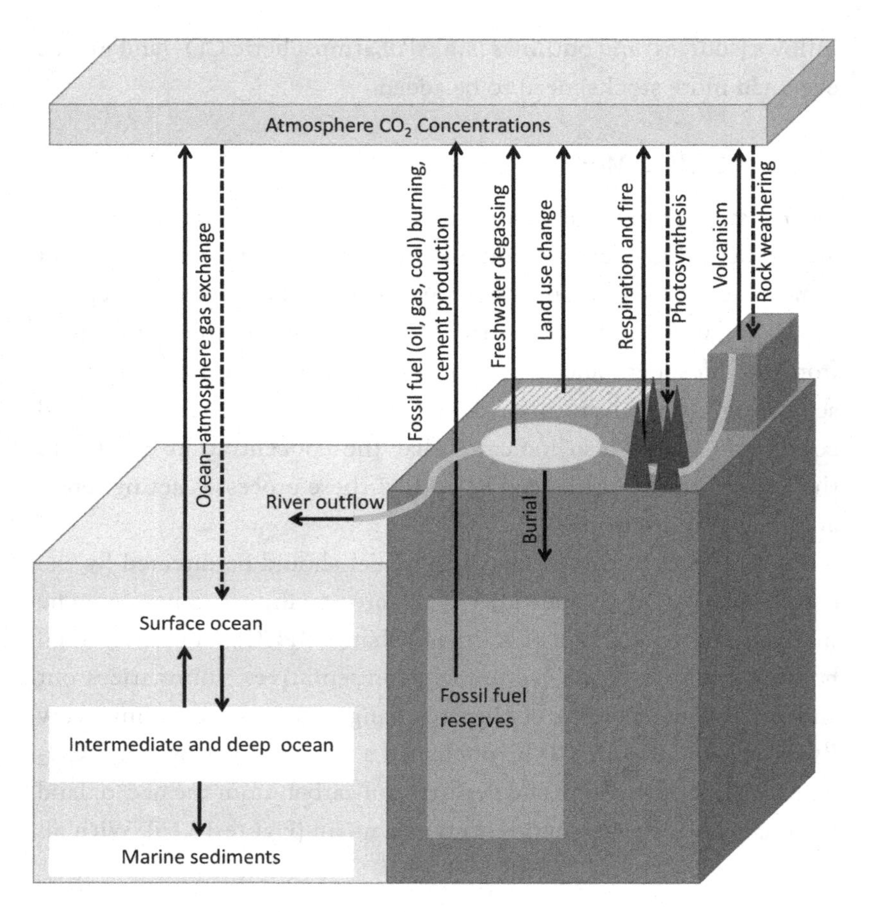

FIGURE 12.15 The global carbon cycle.
Adapted from Ciais et al. (2013).

of these major carbon reservoirs and flows linked through a model, the big picture changes in atmospheric CO_2 can be captured.

12.5 TAKE-HOME MESSAGES

- We use models to study all kinds of environmental systems, including dynamic systems.
- Models (conceptual and quantitative) are simplified representations of a system.

- For a mathematical model to work, the mathematics must work.
- A linear model relates one variable (the response) to one or more other variables (predictors).

REFERENCES

Ciais, P., C. Sabine, G. Bala et al. (2013). Carbon and Other Biogeochemical Cycles. In: *Climate Change 2013: The Physical Science Basis. Contribution of Working Group I to the Fifth Assessment Report of the Intergovernmental Panel on Climate Change*, T. F. Stocker, D. Qin, G.-K. Plattner et al. (eds.). Cambridge University Press, Cambridge, UK, and New York.

Keeling, C. D., S. C. Piper, R. B. Bacastow et al. (2001). Exchanges of atmospheric CO_2 and $13CO_2$ with the terrestrial biosphere and oceans from 1978 to 2000. I. Global aspects, SIO Reference Series, No. 01-06, Scripps Institution of Oceanography, San Diego, CA, 88pp.

US Department of Energy, Carbon Dioxide Information Analysis Centre. (2018). http://cdiac.ess-dive.lbl.gov/trends/emis/meth_reg.html. Last accessed September 4, 2018.

PART III Communicating Environmental Science

Chapter 13

Preparation

1. Before reading Chapter 13, consider the following questions:
   - What should be in a research proposal?
   - How can I find good references?
   - How much detail do I need to include in the methods section?
   - How can I avoid plagiarism?
2. Below is a list of things that have to happen in order to make a pizza. Number the steps in the order that they should happen. Draw a flow chart that depicts the sequence of events. If two things can happen at the same time, consider splitting your flow chart into two streams to show more than one activity occurring at the same time.
   - Preheat the oven to a high temperature.
   - Buy flour, yeast, tomato sauce, cheese, and whatever else is needed to make a good pizza.
   - Grate the cheese.
   - Roll the dough into a flat circle.
   - Mix warm water and yeast; let sit for 15 minutes.
   - Cook the pizza for about 15 minutes.
   - Add the flour to the yeast mixture, knead for 5 minutes.
   - Let the dough sit until doubled in size, about 1.5 hours.
   - Invite friends over.
   - Spread the tomato sauce, cheese, and other ingredients on the dough.
   - Then let the pizza cool a few minutes before cutting it into slices.
   - Enjoy eating the pizza with your friends.
   - Prepare the table / find plates and napkins.

13 Writing a Science Proposal

13.1 THE STRUCTURE OF A PROPOSAL

Writing a proposal is the first step to getting a project approved and funded. In many cases, a call for proposals is like a competition where the most persuasive proposal will get approved and others will not. In science, persuasive writing is not hyperbolic or purposefully evasive. Taking a narrow view of a topic to elevate its importance is not an effective way to write a persuasive science proposal. A persuasive science proposal clearly and accurately articulates the motivating problem and outlines the methodology chosen to address the problem in a logical and systematic way. Scientists use references as supporting documents to authenticate statements. Effective referencing increases the quality of the proposal.

Proposals come in many sizes and formats. When submitting a proposal, make sure you read the instructions provided by the evaluation committee (or assignment guidelines) and follow them closely. Your proposal will be judged on attention to detail, demonstrated by close alignment with the instructions. If few instructions are given, remember that to write a persuasive proposal, you have to clearly and concisely communicate:

- the fundamentals of the project, i.e., the research questions and methodology;
- the relevance of the project, i.e., why the project should happen now; and
- the implications of the project, i.e., what could happen as a result of your proposed work.

Sometimes a proposal structure will be prescribed. But if you have to write a proposal without a prescribed structure, you can use a typical structure with the following sections:

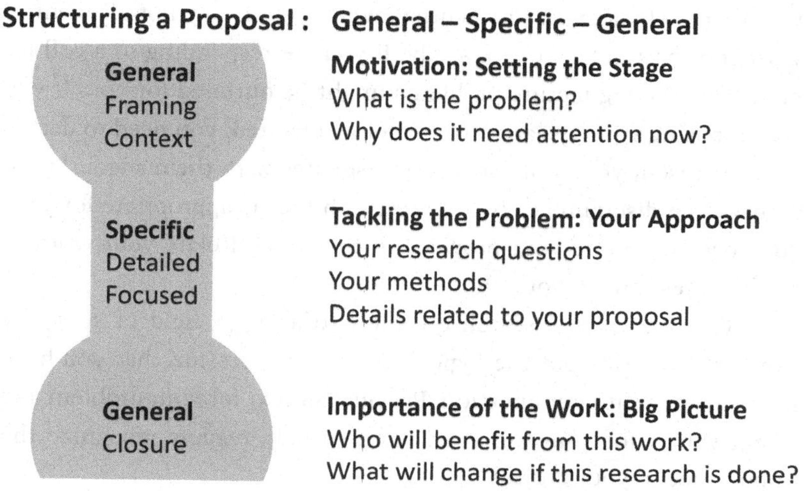

Structuring a Proposal : General – Specific – General

General
Framing
Context

Motivation: Setting the Stage
What is the problem?
Why does it need attention now?

Specific
Detailed
Focused

Tackling the Problem: Your Approach
Your research questions
Your methods
Details related to your proposal

General
Closure

Importance of the Work: Big Picture
Who will benefit from this work?
What will change if this research is done?

FIGURE 13.1 Schematic representing the structure of a research proposal. Begin with general information, then provide concrete details, and end with a connection to the big picture.

1. Background / Introduction
2. Relevance
3. Your research question(s)
4. Methodology
5. Timeline
6. Budget
7. Implications
8. References

A proposal moves from general information to specific information and back to general again: it has the shape of an hourglass (Figure 13.1). The hourglass metaphor reminds you to provide context, then describe the critical details about your project, before you leave readers with the big picture importance of your work.

13.1.1 Background / Introduction: The Motivation of the Work

To communicate the importance of your work, you need to know who your readers are and what is important to them. A science proposal is usually read by a panel of scientists who may or may not all identify with your particular field of study. If you are responding to a call for proposals, the big-picture problems might be outlined for you. If you are proposing your research to a panel unsolicited, you need to decide how to present your work so that it resonates with the readers. Try to identify the discipline or field of research that is appropriate for your audience and frame your proposal in that field. Relate your work to the big questions in your chosen field.

Connect your research question to a larger field of study by explicitly pointing out the logical thought progression that you have made to arrive at your question. Provide context for your problem and a motivation for your proposed work to help readers recognize the importance of your project.

Throughout your proposal, the references that you include send signals to the readers. In the context section, your references indicate how well you understand what is important in your field.

- Are you choosing to cite foundational work (probably older and highly cited papers) or new cutting-edge ideas (newer papers in prominent journals)?
- Are all your references from the same research group or are you including different perspectives on the topic?

Aim to include critical papers by the key researchers in your chosen field. If possible, cite a variety of papers from different research groups. This demonstrates that you have presented a broad background of the issue. Take some time to think about your references and the messages that they convey. Make sure you are sending the signals that you want to send.

13.1.2 Relevance: Establishing the Need to Act

Once you have established a connection between your research question and the larger problems in your field of study, you still need to explain why your research should be supported now. The "Why now?" messaging in a proposal is called the relevance statement.

Explicitly state why your research is timely. Relevance statements commonly express

- a new opportunity: Is an event about to take place, or do we now have the technology to answer this question, or have samples been obtained from a new location?
- a new challenge or recently identified gap in knowledge: Are you addressing a critical yet unanswered question in your field?
- a new approach to a known problem: Are you suggesting a new interdisciplinary methodology? Are you refining or testing an already established idea?

13.1.3 Research Questions

The research question is the heart of a proposal and directs the proposed work. Therefore, the research question should clearly and explicitly link to the work that you are proposing. Don't assume that the question will be understood from implied language in your introduction. The questions should state the "what, where, when and how" of your study (see Chapter 5). Your question(s) should be specific and detailed enough for the readers to follow the logical connections that you are making between the motivation for the work and the work itself. An elegant writer will connect the context and relevance so that the research questions that follow feel like the natural questions to ask, given the challenge faced or opportunity provided.

13.1.4 Methods

The methods that you propose demonstrate your understanding of the problem, your familiarity with the field, and your creativity. The resolution of your data collection, measurements, or analysis must capture the variability of interest in the system or phenomenon that you are studying. There is tension between data collection / analysis and time / money. You need to find a balance that will provide you with the information you need in a reasonable timeframe and budget. The science should come first, but make sure you are being efficient with your time and resources.

Communicate your proposed methods to highlight your understanding of the methods, including the caveats and limitations of the methodology you have chosen. You do not need to provide all of the detail of your analysis: do not write a sampling or analytical protocol. Instead, the proposal methods should indicate that you know what is important to communicate and what is extraneous detail.

Important things to include:

- the location of field work;
- the type of samples that will be collected (soil, air, water, species etc.);
- the sampling resolution (in time and space);
- the way samples will be collected, stored, processed, and analyzed (balance between not enough and too much detail);
- the techniques used to make the measurements (inductively coupled plasma-mass spectrometry (ICP-MS), high performance liquid chromatography (HPLC));
- the caveats to your methods and how they are addressed; and
- the data used from other sources.

13.1.5 Timeline / Budget

The timeline and budget indicate how much you have thought through your project. The reviewer will be checking if your project can actually be done in the timeframe stated and with the amount of money requested.

Though the details of the timeline and budget might best be presented using tables, it is important to introduce the key features of the project timeline and budget in a short descriptive paragraph. In addition to the total cost of the project, the budget table should include the major line items associated with your methods and how much each would cost.

To convince your reviewers that your project is feasible, outline the major steps in your methods and roughly when they occur in time. Find a clear and succinct way to capture critical information. Timelines might need to show multiple things happening at the

	Apr	May	Jun	Jul	Aug	Sep	Oct	Nov	Dec
Field site preparation									
activity 1									
activity 2									
activity 3									
Sample collection and analysis									
activity 4									
activity 5									
Data analysis									
activity 6									
Writing									
activity 7									
activity 8									

FIGURE 13.2 Example of a project timeline displayed as a Gantt chart. The time steps are presented as columns; the activities are presented as rows. Shaded areas depict the duration of the specific activity.

same time. A Gantt chart (Figure 13.2), a spreadsheet showing time as columns and activities as rows, is usually quite effective.

If you are using tables and/or figures to present your timeline and budget, remember that you need to cite the tables and/or figures somewhere in the text of your proposal.

13.1.6 Implications

The implications section of your proposal is where you articulate the possible future effects of your work. What academic theories will your work contribute to? What specific future research will it enable or guide? What real-world decisions might be informed by your work?

When discussing the implications of your work, do not presuppose your results. You need to explain why the proposed project is worthy, not just one possible anticipated result. This is where an objective approach is important. Your research should be useful, whatever results are found.

13.2 MOVING FROM THINKING TO DOING: SCOPING, PLANNING, AND PROPOSING A PROJECT

Imagine you want to throw a birthday party. If it is not spontaneous, you likely go through a planning stage. You might decide how many people to invite and why. You might think about how they would interact and then decide whether this is a house party with open attendance or a dinner party with a particular guest list. You will need to decide what food will be served. Is this party a potluck or are you providing everything? What recipes will you use? What drinks will be served? When will you do the shopping? Can anything be made in advance? Do you have enough plates and forks and glasses? For the party to be successful and fulfil your intentions, planning the party is important.

When planning a science project, you need to go through this same process. The first thing you need to do is decide what the project is exactly – and what it is not. This decision stage is the scoping stage of a project. Scoping a project helps you understand the field of study that you want to contribute to, the timeframe of the project, and the end goal of the project. Like deciding between a house party and a dinner party, scoping allows you to set expectations. Once you have decided on the scope of the project, then you can start planning the details. To help you make decisions about the methods, the scale, and the data required, you should consult published peer-reviewed papers. Thoughtful decisions early on will help the project be successful.

A science proposal is the way you communicate your project plan to others – particularly to others who might be in the position to help you make your plan a reality: a potential supervisor or a funding agency. Reviewers are convinced, not through exaggerated language or overstated claims, but through a distilled articulation of the problem (demonstrated by referencing published work), including the question you propose to address and the methods you will use to answer the question well. The reviewers will also want to know the importance of your proposed work in terms of what will be improved or what will change as a result of your work.

This is challenging to pull off. You might have a great idea that is timely and will generate important change, but if you can't communicate your idea, your proposal will not be successful. Effective communication is based on good ideas that are well organized and clearly written. Don't expect to be able to write a great proposal in one draft. You will need to iterate to improve your ideas, organization, and writing.

13.2.1 Step One: Building an Annotated Bibliography and Outline

13.2.1.1 Organization

An outline is a way to organize your ideas. The focus of the outline stage is to identify the key points that you want to cover in your proposal and to decide the order in which you will present critical information. The outline stage is also the time to gather supporting documents and peer-reviewed references. In your outline you should use in-text citations to ensure that you have references for all of the major points that you will present in your proposal. Building an outline allows you to organize your ideas and gather targeted supporting evidence efficiently.

Decisions about the content of your proposal and the organization of the content need to happen at different levels: macro- (heading / section), meso- (subheading / subsection), and micro- (paragraph) levels. An outline helps you focus on the macro and meso scales, or the headings and subheadings of your proposal. Once you have determined the structure of your outline, you can start thinking about the content that you will need to develop for each section.

13.2.1.2 Finding Appropriate References

To be scientifically credible, you will need to support your argument (question and approach to answering the question) with previously published peer-reviewed science (see Box 13.1 and Chapter 1).

BOX 13.1 Citing and Referencing Scientific Articles

In scientific writing, citations are used to refer readers to the source of unique, specific information that is not likely known by the general population. Citations directly link new scientific work to past scientific work. In this way science builds one idea on another.

The citation should link directly to the original source of an idea or piece of information. In a scientific paper, the purpose of the introduction is to provide context for the new work; therefore, the introduction usually summarizes previous work. It might be tempting to cite a paper for ideas presented in the introduction, but these ideas likely originated somewhere else. A diligent writer will follow the paper trail, citation by citation, to find the original source of the idea and properly cite the original article.

Different disciplines have different conventions on how to cite information. In science writing, there are two dominant approaches: listing the author and year in parenthesis like this (last name, year) or using a superscript numbering system like this.[1]

These two approaches have different strengths. Using the (last name, year) convention allows informed readers to recognize the author's name and see the date of publication and evaluate the relevance of the citation as they are reading the information. The citation itself is, therefore, a critical part of the logical argument presented by the authors. Conversely, the numbering system allows readers to absorb a paper and new information without disruption. The text flows more easily; sentences are not interrupted with diverting information. Every journal will outline the expectation for authors, including how citations should be executed. When submitting a paper, check the guidelines for citations and make sure to use the identified convention. Within one paper, always be consistent with your citation format.

All of the information presented in a new scientific paper must be in the words of the new author. Information presented that originated elsewhere, including in previously published scientific articles, must be paraphrased. Direct quotations, even if placed in quotation marks and cited, are not used in scientific writing. Within a sentence that

BOX 13.1 **(cont.)**

includes paraphrased information, citations can be placed either immediately following the specific information or at the end of the sentence (before the period).

The citation is a short-hand notation of the reference. The full reference list (compiled at the end of your text) is a list of all of the sources cited in the text. In order for a source to be on the reference list, it must be cited in the text. This is different from a bibliography, which is a complete list of all sources that have contributed to your understanding of a topic and that shaped your writing. A bibliography can contain sources that might not show up as a citation in your text. In science writing, citations and a reference list, not a bibliography, are used to support the ideas that are being presented.

Each peer-reviewed paper is a unique contribution to the body of scientific knowledge. Each paper represents a rigorous question-driven process of data collection and analysis from which conclusions are drawn. In science we cite a paper either to show the original source of an idea, or to show that a method we are employing has been tested and used previously. When looking for references, focus on finding papers that you will cite because of information presented in the methods, results, discussion, or conclusions of the paper, not the introduction. The introduction section sets the stage, or prepares the readers, for new scientific work: the introduction is a summary of previous work, not the new work that is presented in the paper. If you are intrigued by content in the introduction of a paper, follow the citation string backwards to find the original source of the intriguing idea.

The process of finding good references is a puzzle in itself. You will want a balance of references that demonstrate that you know the critically important concepts in your field and the cutting-edge ideas that are still to be developed. It takes time to search the published literature. These steps should help you organize your search:

1. Decide what database to use: choose a curated database of peer-reviewed content (see Chapter 6).
2. Choose search terms that are specific to your big-picture problem.
3. Refine the search, using the database features, to focus on your discipline: limit the search to journals specific to your field.
4. Arrange the papers first in order of citations (what papers have been cited the most); consider collecting the top papers here if they are appropriate to your topic.
5. Rearrange the list from youngest to oldest in order of publication date; consider collecting a few newer papers if appropriate.
6. Identify the journals that you now have represented in your list of publications. Are all papers from the same journal? Do you have some references from specialist journals (highly technical) and some from generalist journals (cutting-edge, cross-cutting science of interest to many fields)?
7. Look at the keywords listed in the papers that you think are most aligned with your ideas. Note if there are keywords used that you did not include in your original search.
8. Change your search terms and go through this process again. If you get new papers, repeat 4, 5, and 6.
9. Skim the references of your chosen papers (this can be done in the database); maybe you can find a series of papers that relate to your topic or that develop an idea over time.
10. Compare your references to your general outline. Do you have enough papers to put appropriate citations in each section? Are you relying very heavily on one or two papers (and the rest are filler) or do you place an equal weight of importance on at least half of your papers? Aim for the latter. If you are relying on a paper for information that is in the introduction, find the original citation of the idea.

13.2.1.3 Building and Using an Annotated Bibliography

Building an annotated bibliography, a list of papers with associated notes, is an extremely useful, focused, and efficient way to capture information or ideas or from a reference that are useful for your particular purpose. Rather than taking random notes, decide what aspect of the paper is useful to you. Is it the methods or approach used? Is it the conclusions drawn? Then, extract the information that you need.

Take detailed notes, capture quantitative, specific information but **do not write down the sentences verbatim: Do not copy and paste paragraphs.**

This ensures that when you write your proposal, you will use your own words and not fall into the trap of thinking that the original authors "said it best."

With your annotated bibliography in hand, you are now prepared to flesh out your outline by adding the important ideas (supported by citations) that you will discuss in each section.

13.2.1.4 Finalizing the Outline

Make a list of the compelling arguments that you will include in each section of your proposal to concisely communicate the importance of your research question, the feasibility of your approach, and the benefits (both academic and societal) of completing this work. Recall that an argument in science is supported by evidence – the more concrete the better. When structuring your argument, make sure you include and cite the evidence in support of your claim.

The final outline stage is where you will start to think about the reader. It is likely that your readers are scientifically literate but not experts in your field. Therefore, you need to ensure that you present the elements of your argument in an order that your readers can follow. Try not to presuppose knowledge or refer to information that the readers will get just a little bit later. Your final outline should be the skeleton of your full proposal, including all of the important components of your argument laid out in a systematic way that logically builds from one idea to the next.

13.2.2 Step Two: Fleshing Out the Proposal

To transform your outline into a full proposal, you will need to focus on the meso- and micro-structure of the proposal. Within each section (or subsection) of the proposal, you will likely have more than one main idea that you are trying to convey. Therefore you will need more than one paragraph for each section.

A paragraph is a unit of writing that communicates one main idea. Each paragraph should concisely convey a well-crafted message. The first sentence of the paragraph should state the main idea that will be discussed and expanded upon in the paragraph. The length of the paragraph should be dictated by the content required to convey the specific topic of the paragraph. This could be as little as two sentences but is usually more. The last sentence of the paragraph should be the intended take-home message. Start a new paragraph when you initiate a new idea or change your focus.

13.2.3 Step Three: Editing

After you have finished your first attempt at putting your ideas on paper, take a break: one hour is good; a day or two is better. When you revisit your proposal, ask yourself the following:

- Is your research question focused and specific? Does it indicate the *what, where, when,* and *how* of your study (see Chapter 5)?
- Have you motivated your particular research question? Is the context you provide targeted towards the problem you are investigating? Have you included any irrelevant or unnecessary information?
- Have you provided a clear case for why your work should happen now?
- Are your methods aligned with your research questions? Will you be able to answer your question by doing what you have described?
- Did you base your implications on one anticipated outcome rather than provide a general argument outlining the need for your results?

When you revisit your proposal, it is time to be brutal and edit. Well-structured paragraphs and well-written sentences make the difference between a great idea that falls flat and a funded proposal. When editing, aim to remove 10% (or more) of your words. This time

- Really read your paragraph topic sentences. Make sure all of the information in the paragraph is there: move it if it should be somewhere else; delete it if you have said it before; or make a new paragraph if you drift onto a new topic.

- Consider your word choices. Aim to use specific, clear words (see Chapter 5). If a word is the right word for your purpose, use it. Do not change a word just for style. The reader will think you are being nuanced and get confused. Delete redundant words and filler phrases. Replace ambiguous words, and define jargon.
- Focus on your transitions. Are they necessary or just filler? Check that they make sense and make sure they link important ideas. Delete them if they do not add information.
- Make sure your information is correct and cited. All tables and figures should be referenced in the text in the order that they appear.
- Don't rely on old papers for current information. Double-check the publication year of the papers you are using to make sure that your information is up to date.
- Finalize your references. All references should be cited somewhere in the text. All references should be complete and formatted consistently.

Before you submit your proposal, recheck any guidelines you have been given. A proposal can be rejected if it does not meet the guidelines.

13.3 TAKE-HOME MESSAGES

- Proposals are submitted to get a new project approved and funded.
- In addition to the research question and methodology, proposals need to state the relevance and implications of the proposed work.
- A research proposal is a plan of action and needs to include enough, and only enough, information to demonstrate that you could successfully complete the proposed work.
- Writing a proposal takes time; you will need to outline your argument, find supporting references, and construct your sections, paragraphs and sentences thoughtfully.
- When writing a proposal, make sure you follow any instructions provided.

Chapter 14

Preparation

1. Before reading Chapter 14, consider the following questions:
   - What is the purpose of an abstract?
   - What is the difference between an abstract and an introduction?
   - Who reads the abstract of a paper and who reads the introduction?
2. A typical science paper is organized into sections: Introduction, Methods, Results, Discussion, Conclusions. Draw a diagram that relates the content in the abstract to the **content** in these typical sections. This doesn't mean place the abstract in order (the abstract comes first). It means connect the information that is presented in these typical sections to the information that is presented in the abstract.
3. Consider your research proposal or a topic that you find interesting. Jot down the information that you would include in the abstract and the information that you would include in the introduction.

14 Writing an Abstract

Scientific writing is commonly preceded by an abstract: a stand-alone paragraph that presents the key points of the paper.

⊗ An abstract is not an introductory paragraph.
⊗ An abstract is not an outline of what will be discussed in the paper.
⊗ An abstract is not a chronology of the work that has been done.

An abstract is a summary or precis with a particular focus to communicate new scientific findings in context. Reading the abstract is not a substitute for reading the full paper, but the abstract should communicate the keys points that will be further elaborated in the paper.

In preparation for Chapter 1, you read the following abstract:

A Global Map of Human Impact on Marine Ecosystems
The management and conservation of the world's oceans require synthesis of spatial data on the distribution and intensity of human activities and the overlap of their impacts on marine ecosystems. We developed an ecosystem-specific, multiscale spatial model to synthesize 17 global data sets of anthropogenic drivers of ecological change for 20 marine ecosystems. Our analysis indicates that no area is unaffected by human influence and that a large fraction (41%) is strongly affected by multiple drivers. However, large areas of relatively little human impact remain, particularly near the poles. The analytical process and resulting maps provide flexible tools for regional and global efforts to allocate conservation resources; to implement ecosystem-based management; and to inform marine spatial planning, education, and basic research.

(Halpern et al., 2008)

This is the introductory paragraph from the same paper (Halpern et al., 2008):

> Humans depend on ocean ecosystems for important and valuable goods and services, but human use has also altered the oceans through direct and indirect means [5 references provided]. Land-based activities affect the runoff of pollutants and nutrients into coastal waters [2 references provided] and remove, alter, or destroy natural habitat. Ocean-based activities extract resources, add pollution, and change species composition [1 reference provided]. These human activities vary in their intensity of impact on the ecological condition of communities [1 reference provided] and in their spatial distribution across the seascape. Understanding and quantifying, i.e., mapping, the spatial distribution of human impacts is needed for the evaluation of trade-offs [or compatibility] between human uses of the oceans and protection of ecosystems and the services they provide [3 references provided]. Such mapping will help improve and rationalize spatial management of human activities [1 reference provided].

Dissecting these two paragraphs helps differentiate between an abstract and an introduction. In this example, the abstract concisely presents the argument of the paper including the context, the objective, and the new evidence (or results) that demonstrate that the objective has been met.

An abstract typically outlines:

1. the motivation,
2. the research question,
3. the methods,
4. the key results or conclusions, and
5. the implications of the presented work.

The authors of this abstract (Halpern et al., 2008) use key words to concisely communicate four of these five elements:

Motivation: The management and conservation of the world's oceans **require** ...
Methods: We **developed** ...
Results / conclusions: Our analysis **indicates** ...
Implications: The analytical process and resulting maps **provide** flexible tools for ...

The one element not directly stated is the research question. In this abstract, a research objective (not phrased as a question) is implied by the title: *A Global Map of Human Impact on Marine Ecosystems*. If you reword the title as a question, the objective becomes clearer:

Can we generate a global map of human impact on marine ecosystems?

or

What has been the human impact on marine ecosystems?

The objective of the authors is also outlined as a need in the first sentence: "The management and conservation of the world's oceans require *synthesis of spatial data on the distribution and intensity of human activities and the overlap of their impacts on marine ecosystems*" (Halpern et al., 2008).

It is common for the research question to be implied rather than stated. However, this can lead to ambiguity. A clear research question drives the logic of the writing. Without a clearly stated research question or objective, it is possible to stray without noticing. When writing your abstracts, consider stating your research question or objective directly.

In contrast to the abstract, the introduction introduces and elaborates on the motivation that drives the work. The introduction provides context, including citations, to support the logic of the objective. Structurally, the introduction is the beginning of the paper. Therefore, the introduction should start from the beginning. Do not get mislead by the position of the abstract at the top or front of the paper. The abstract is an additional element to the paper; the introduction is where the paper starts.

In the example above, the authors clearly state the problem in the introduction. They then elaborate on the problem, providing details that allow the readers to see the extent of the challenge. They motivate their research question by indicating what might help solve the problem and they reinforce their approach by stating a use to which their research output could be put:

Statement of the problem: "Humans depend on ocean ecosystems for important and valuable goods and services, but human use has also altered the oceans through direct and indirect means" (Halpern et al., 2008)

Elaboration: "Land-based activities affect the runoff of pollutants and nutrients into coastal waters and remove, alter, or destroy natural habitat. Ocean-based activities extract resources, add pollution, and change species composition. These human activities vary in their intensity of impact on the ecological condition of communities and in their spatial distribution across the seascape" (Halpern et al., 2008).

Suggested solution: "Understanding and quantifying, i.e., mapping, the spatial distribution of human impacts is needed for the evaluation of trade-offs (or compatibility) between human uses of the oceans and protection of ecosystems and the services they provide" (Halpern et al., 2008).

Use for the research / implications: "Such mapping will help improve and rationalize spatial management of human activities" (Halpern et al., 2008).

A total of 10 peer-reviewed papers are cited in the introductory paragraph of Halpern et al., (2008). The papers cited help to validate the motivation and logic of the new paper based on past science. The introduction relies on past peer-reviewed work to provide the context for the new proposed work. (See Chapter 13 for a discussion on including references in your introduction.) In contrast, the abstract is focused specifically on the key ideas presented in the new paper that follows. Therefore, abstracts rarely include citations.

14.1 TAKE-HOME MESSAGES

- An abstract is a scientific summary that communicates the motivation, the research question, the methods, the key results or conclusions, and the implications of the presented work.
- An abstract is a stand-alone paragraph that doesn't take the place of the introduction.
- An abstract commonly does not include citations as the ideas presented in the abstract are from the new paper that follows.

With your abstract complete, you are now ready to submit your proposal.

Good luck.

REFERENCE

Halpern, B. S., S. Walbridge, K. A. Selkoe et al. (2008). A global map of human impact on marine ecosystems. *Science*, 319 (5865): 948–952.

Epilogue

E.1 EVALUATING A RESEARCH PROPOSAL

This book is both a resource and a practical guide to thinking about, doing, and communicating science. After you work through concrete examples of how to develop a scientific question, how to consider the complexity of natural phenomena, and how to align questions with data analysis, a scientific research proposal is used to demonstrate the degree to which critical environmental science concepts have been absorbed and applied. As an assignment, a research proposal is an effective way to integrate core concepts of scientific thinking while allowing students to engage with a topic of particular personal interest.

Typically, a research proposal is evaluated by a funding agency or a research panel. In a classroom, evaluating a research proposal might happen as part of a course assessment. Whoever is evaluating the research proposal, they will try to determine

- if the science being proposed is interesting, new, and useful;
- if the writer is demonstrating insight of the problem;
- if the research plan will generate an answer to the question;
- if the research plan is practical enough to be executed; and
- if there is a reason to support this project before / instead of another.

Funding agencies will use unique criteria to evaluate proposals, including how well the proposal incorporates strategic initiatives or themes. Within a classroom, the evaluation criteria can be developed and communicated to students in a way that outlines the features of different assessment criteria that would place a proposal into a gradient of categories leading up to exemplary.

E.2 ASSESSING A RESEARCH PROPOSAL USING A RUBRIC

A rubric is a tool used to both communicate assignment expectations and to assign grades in a systematic way that includes feedback. The rubric below outlines in detail the required elements of a research proposal broken down into three categories: framing, proposed research, and feasibility. Making this rubric available to students as they work on a proposal assignment allows them to understand important aspects of each criteria and to self-evaluate through the process of writing.

E.3 EXAMPLE RUBRIC FOR A SCIENTIFIC RESEARCH PROPOSAL

The objective of this assignment is to develop and hone your abilities as an independent researcher of environmental science. In particular, you will review and synthesize a body of literature in an area that you choose, identify a compelling research question that follows from this review, and propose a means for addressing your question through your own hypothetical future research.

A research proposal is a piece of persuasive writing. The proposal should convince your reviewers that your research will answer a critical question that needs to be answered now. Reviewers are convinced, not through hyperbolic language or overstated claims, but through a distilled articulation of (1) the problem (demonstrated by referencing published work) (2) the question you propose to address the problem, and (3) the methods you will use to answer the question. The reviewers will also want to know the importance of this proposed work in terms of what will be improved or what will change as a result of your work.

E.3.1 Framing: Title and Abstract / Summary

The title is the first articulation of the information being presented. It should clearly represent the goal of the proposal. An abstract is a concise summary of the proposal including the context / background

6 marks	**Emerging** 0–60%	**Developing** 61–80%	**Mastering** 81–100%
Title (0.5)	Does not convey the content of the proposal.	Readers can tell topic by reading title. Too long; wordy; too short; focus is unclear.	Informative, concise, and engaging. Reflective of the content of the proposal.
Abstract (1)	Vague, does not articulate the key elements of the proposal.	Presents the motivation for the research, the questions, and the approach. May contain a few extraneous points. Relevance or implications might be unclear.	Clearly and concisely captures the essence of the proposed research and its importance. Each sentence is useful and informative.
Context (2.5)	The general area of inquiry and / or the real-world challenges are unclear. Background context lacking.	The general area of inquiry, and real-world challenges are clear. Academic approaches, theories, and contributions are included but might not be clearly relevant.	The general area of inquiry and real-world challenges are clear. Relevant academic approaches, theories, and contributions are included.
Presentation (2)	Writing errors detract from comprehension.	Writing is wordy, ambiguous, or jargon laden.	Writing is clear and concise.
Comments / questions arising			

motivating the research, the research questions, the approach proposed, and the implications of the work. In your submission, choose the correct label for your approach.

E.3.2 Proposed Research

Articulate a clear research question or questions. Explain why your research can or should happen now. Support your statements by citing academic literature and real-world need. Clearly explain why your chosen question is timely and important to the field and to society more broadly.

Outline the methods that you propose to use to answer your research questions. The methods should indicate the approach you will take and demonstrate enough familiarity that it is clear you know what you are doing. You should not get bogged down with detail. (The methods are not intended to be a specific sampling protocol.) Use references to support your choice of methods.

E.3.3 Project Planning / Feasibility

The budget and timeline indicate how much you have thought through your project and whether it is, in fact, feasible. The reviewer will be checking to see if this project can actually be done in the timeframe stated and with the amount of money requested. To convince your reviewer that your project is feasible, outline the major steps in your methods and roughly when they occur in time. Include a budget that identifies the major line items associated with your methods and approximately how much each would cost. You can assume that you will have access to certain infrastructure and materials from your own sources or those of your supervisor, but be explicit about these assumptions. (E.g., you might state that you'll need the use of a field vehicle, and you might assume that your supervisor will have one; you might, therefore, request from funders only the cost of fuel and maintenance.) If you are planning sample analysis, try to find a reasonable estimate of the costs (check commercial labs for example costs of chemical analysis) and calculate your costs based on the number of samples you expect to run. We don't expect formal cost

11 marks	**Emerging** 0–60%	**Developing** 61–80%	**Mastering** 81–100%
Research questions (2)	Questions cannot be tested.	Questions are too broad or too vague to be tested well.	Question is clear and testable. Question is specific and concise answering "what, where, when, and how" in relation to the problem.
Relevance (2)	Proposal importance is not adequately established. The historical research presented does not justify the proposal.	The importance of the proposed research is stated but is not well supported by past work, defined opportunities, or society's needs.	The specific need for the answers to your particular research questions is clear and referenced well.
Methods (5)	Methods are not clear. Methods do not address the research question adequately. Method steps are not detailed enough to demonstrate an understanding of the research question.	Methods address the research questions. Methods are detailed enough to address the research questions suitably. Method steps are systematic, clear, and easy to follow. Extraneous information might be present. References	Methods are clear, concise, easy to follow, and systematic. The methods are specifically written to answer the research question(s) and do not contain extraneous material. Method steps demonstrate the breadth of the proposed question(s). References

(*cont.*)

11 marks	**Emerging** 0–60%	**Developing** 61–80%	**Mastering** 81–100%
		supporting the methods are included.	supporting the methods are included.
Implications (2)	Implications are absent or do not align with the proposed research. Results are presupposed and implications are written only for the desired results.	Implications are vague or align only indirectly with the proposed research. Implications might deviate from the context provided.	Implications are clearly articulated, align with the research presented, and follow from the context and relevance.
Comments / questions arising			

estimates or precise times. Make sure you include the total cost as well as the line items of the budget.

If you are using this rubric as an instructor to provide feedback on a class assignment, consider also having student peers evaluate each other's work. The process of reading and evaluating another example of the same assignment provides an opportunity for students to gain a more objective view of their own work.

If you are using this rubric as a student conducting a self-evaluation, read the full descriptions of each criterium, not just the Mastering column, to identify elements that you might not have considered during your writing process. Iteration always improves your work.

3 marks	Emerging 0–60%	Developing 61–80%	Mastering 81–100%
Timeline (1)	Timeline is not very realistic or detailed.	Timeline represents the variability of the system appropriately. Most significant milestones are captured.	Timeline captures the variability of the system and indicates alignment between the phenomena studied and the data collection. The timeline is presented clearly and effectively.
Budget (1)	The budget is absent, incomplete, or missing major elements related to the proposal.	The budget captures the main elements of the proposed research. Some extraneous detail may be present.	The budget captures all critical components of the research. The budget is presented clearly and effectively.
Figures, tables (0.5)	Figures and/or tables are a distraction. They add no meaningful content.	Figures and/or tables are clearly relevant and mostly align with the text and the reasons presented. Figure numbering and/or captions, and/or citations are incomplete, missing, out of order, or inappropriate.	Figures and/or tables are clearly relevant, align with text, and illustrate key points. All figures and tables are clearly numbered and cited in the correct order. All figures and tables include informative captions.

(*cont.*)

3 marks	**Emerging** 0–60%	**Developing** 61–80%	**Mastering** 81–100%
References and citations (0.5)	Few to no citations in the text. Incomplete citations, and/or too few references, and/or several references appear to be irrelevant to topic, and/or most references are from non-peer-reviewed sources. Citations are not in specified format.	Citations mostly align with reference list. Full citations listed in specified format. Most references listed appear to be relevant to topic at hand. Some references are not from peer-reviewed sources.	Citations align with reference list. Full citations listed in specified format. All references listed appear to be relevant to topic at hand. All references appear to be from peer-reviewed journals (except in exceptional circumstances).
Comments / questions arising			

As the field of environmental science engages so intimately with communities and societies, communicating science is emerging as a critical skill for environmental scientists. Introducing open-ended problems, such as proposing new science, is an effective way to provide students with an authentic scientific experience in the classroom. Evaluating new ideas is not only critical to science, it is critical to society. Using a scientific proposal to have students practice *thinking* scientifically, *doing* basic data processing, and *communicating* science effectively will prepare students to engage actively as scientists and as citizens.

Appendix: Working in Excel

There are many online resources that explain how to execute calculations in Excel. If you are new to Excel, I suggest that you take a few minutes to learn how to write equations and how to work effectively using Excel spreadsheets.

Excel is a spreadsheet-based application that allows numbers, data sets, and numeric series to be organized in rows and columns on a "sheet" for the purposes of executing calculations and generating graphs. Within one Excel file, or workbook, you can add as many sheets as you want. Sheets are typically identified by "tabs" visible at the bottom of the workbook. If you add sheets within a workbook, you should rename them from sheet 1, sheet 2, etc. to appropriately represent the content of the sheet. When starting to work with data, make sure you keep a backup copy of the raw data and the metadata that you have on a separate and well-labelled sheet.

Never do any manipulation or analysis on your backup copy.

Each sheet in excel is made up of cells that are organized in columns and rows. Each cell is identified by the column letter (A, B, C...) and the row number (1, 2, 3...). The first cell in the top left corner is cell A1. The cell immediately below cell A1 is cell A2 and the cell to the immediate right of cell A1 is cell B1.

To execute a simple calculation in Excel, you write an equation in the cell where you want the answer of the calculation to be displayed. All Excel equations start with an equals sign (=). Use the following symbols to execute common actions:

+ (addition)
- (subtraction)
/ (division)
* (multiplication)
^ ("to the power of": ^ precedes an exponential)

To enter cell coordinates into an equation, you can either type them in manually or you can click the cell of interest and it will automatically be entered into your equation. If you wanted to add the value in cell A1 to the value in cell A10, you would write

= A1+A10

Use brackets to indicate the order of operations. If you want to add the number in cell A1 to the number in A10 and then divide by the number in B1, you would write

= (A1+A10)/B1

If you are calculating the sum or the average of a number of cells in a single column or row, you can use an array, or series, format. An array is indicated using the format: (first cell:last cell). For example, if you wanted to add up the values in all of the cells between and including cell A1 and cell A10, you would write

= SUM(A1:A10)

If you wanted to count the number of cells in an array between and including cell A1 and cell A10, you would write

= COUNT(A1:A10)

If you wanted to average all of the values in all of the cells between and including cell A1 and cell A10, you would write

= AVERAGE(A1:A10)

You can copy equations from one cell to another. When you do this, the cell coordinates (column letter and row number) in the equation will change but the spatial relationship between cells that was originally designated will remain intact. For example, if you calculate the sum of the array (B2:D2) in cell E2 and then copy the equation from cell E2 into cell E3, the array will automatically update to calculate the sum of the array (B3:D3) because you have moved the equation down one row (Figure A.1).

	A	B	C	D	E	F
1		**Midterm 1**	**Midterm 2**	**Final Exam**	**Average**	
2	**Course 1 Grades**	65	76	82	=AVERAGE(B2:D2)	
3	**Course 2 Grades**	77	64	79	=AVERAGE(B3:D3)	
4	**Course 3 Grades**	81	67	84	77	

FIGURE A.1 Example spreadsheet demonstrating the spatial relationships of cells in Excel. In this example, the equation in cell E2 was copied and then pasted one row below, into the cell E3. Pasting an equation from one cell into another results in an automatic update of the equation. Pasting an equation into a new cell that is one row below, but in the same column as the original cell, will update the equation so that all linked cells will reflect this change. In this example, the new equation in cell E3 refers to the updated array B3:D3.

If you want to copy an equation into a new cell but do not want all or a portion of the original reference cell coordinates to change, place a $ before either the column letter, the row number, or both depending what coordinate you want to remain constant. The $ must be placed before every coordinate that you want to remain constant. In Figure A.1, I copied the cells from column F and pasted them into column G. You can see that when a $ was placed in front of the column letter, no change to the original cell column letter occurred when I copied and pasted the equation. When no $ was used, the cell coordinates all shifted one column to the right, reflecting the fact that I copied and pasted the equation one column to the right. Be sure to use $ for every cell coordinate in an equation that you do not want to change.

After building, copying, and pasting equations, always check to ensure that the equations are, in fact, referring to the correct cells.

Glossary

Abstract: an overview of a scientific article that provides a synopsis of the study motivation, the research question, or goal, the approach taken or methodology used to answer the question, the key results, the conclusions, and the implications of the new information drawn from the study. The abstract is a stand-alone paragraph that does not replace or act as the introduction of the article.

Amplitude: half of the difference between the maximum and the minimum value of an oscillation or cycle.

Anomaly: the deviation of a quantity from an expected, or mean, value. In environmental science, anomaly data sets are derived by subtracting the mean value from the original data set. The resultant anomaly data set reflects the difference between each original datapoint and the original mean value.

Anthropogenic: formed, produced, or caused as a result of human activity.

Averaging window: a defined amount of time over which you are applying an average.

Bar graph: a graphic representation using rectangles (vertical or horizontal) to compare two or more discrete or categorical values.

Bathymetry: the depth of water in a basin. Detailed bathymetry is used to present the depth of the seafloor at various locations.

Bias: in quantitative science, bias is used to describe a weighting of data in such a way as to influence the interpretation of the results. In order for the results to be interpreted well, the bias must be understood, articulated, and accounted for.

Bibliography: a list of books, articles, and resources that were read in the process of producing a book, article, or text of some kind.

Box plot: a graphic presentation that allows multiple distributions to be compared. Data is presented in two boxes, one on top of the other. One box represents the quarter of the data above the mean value and one box represents the quarter of the data below the mean value.

Central tendency: the central or typical value for a probability distribution.

Correlation: a measurement between -1 and 1 that indicates the strength and direction of common variability between two data sets. A positive correlation indicates both data sets' change in the same direction. A negative correlation indicates that the two data sets change in opposite directions.

Dependent variable: in experimental science or modelling, the dependent variable is the output or resulting variable determined by changes in an input (independent variable). The dependent variable is plotted as the *y*-axis on an *x*–*y* graph.

Dormant: not active. In the case of plants, dormant is referent to periods of time when the plant is not photosynthesizing.

Dynamic: a system characterized by constant change. In environmental models this change is usually in time.

Eccentricity: in palaeo-environmental studies, eccentricity describes a cyclic change in the shape of the Earth's orbit around the Sun from more circular to more elliptical and back again. This cycle has a timescale of approximately 100,000 years.

El Niño / Southern Oscillation (ENSO): a coupled ocean and atmospheric phenomenon that varies with a sub-decadal timescale. Changes in the strength of the equatorial Pacific trade winds shift the position of the surface ocean warm water masses and the atmospheric pressure cells. There are three phases of ENSO: El Niño, neutral, and La Niña. El Niño is characterized by weak trade winds, and warmer than normal waters in the eastern equatorial Pacific Ocean. La Niña is characterized by stronger than normal trade winds and warmer than normal waters in the western tropical Pacific Ocean. The Southern Oscillation is a measure of the difference in air pressure between two sites, one in the eastern equatorial Pacific and one in the western equatorial Pacific.

Empirical: based on observation, measurements, or data.

Extrapolate: to extend beyond existing data into an unknown situation with the assumption that the pattern(s) evident in the data will continue in the same vein.

Frequency: the number of cycles, or repeating events, that occur over a particular period of time.

Histogram: a graph that presents the probability distribution of one variable. Histograms might be a series of touching rectangles with each rectangle representing a bin, or subset of the whole. A histogram can also be drawn as a line connecting the centre point of each bin.

Implications: the resulting change or effect that an action, or new knowledge, can have in the future.

In situ: Latin for "on site," meaning in the original place as opposed to being moved. In the example of a measurement, "in situ" refers to a measurement made outside in the environment as opposed to measurements made in a laboratory.

Independent variable: in experimental science or modelling, the independent variable is the known input variable that determines the output (dependent) variable. The independent variable is plotted as the *x*-axis on an *x*–*y* graph.

Indirect measurement: using a substitute measurement to infer information that cannot be collected directly.

Inflow: a term used in stock and flow models to describe the processes that contribute to a stock. An inflow is typically presented as a rate.

Interpolate: to insert a new value into a series by estimating or calculating it from known surrounding values.

Jargon: technical words that are specific to a field of research and likely not known by the general public.

Least squares: see ordinary least squares.

Mean: the average value of a series. The mean is calculated by summing up all of the values in the series and dividing this number by the number of values in the series.

Median: when a series of numbers is arranged in order of magnitude, the median value is the middle value, or, if there is an even number of values in the series, the median value is the average value of the middle two values in the series.

Metadata: the descriptive information that accompanies a data set or spreadsheet. The metadata can provide an explanation of the variables measured or expressed in the data set as well as information about how the data was collected and how it should be referenced.

Methodology: the approach taken to answer a scientific, testable, question.

Mode: the most common value occurring in a series of numbers.

Multidecadal: spanning a timescale of more than one decade.

Non-dimensional: without units.

Normal distribution: a probability distribution arising from the cumulative addition of small amounts of random variability.

Obliquity: the angle of the Earth's axis of rotation in relation to a theoretical axis, perpendicular to the plane of the Earth's orbit.

Operational definition: a usage that is specified or defined for the interpretation of a particular set of data.

Ordinary least squares: a statistical approach used to estimate a linear function by minimizing the sum of the squares of the differences between the observed values for the dependent variable (the *y*-axis variable) and the predicted values that are generated by the calculated linear function.

Outflow: a term used in stock and flow models to describe the processes that reduce a stock. An outflow is typically presented as a rate.

Palaeoclimate: past climate. The term "palaeo" is usually use for periods of time before direct measurements are available.

Pearson's correlation coefficient: a measure of the linear correlation, or relationship, between two variables. Correlation is not a measure of causation.

Photosynthesis: a process used by plants to convert solar energy into chemical energy. During the process of photosynthesis, carbon dioxide is consumed by the organism and oxygen is produced as a byproduct.

Precession: the circular drift, or wobble, of the Earth's axis. Currently North, as defined by the Earth's axis, points toward the star Sirius. The precessional "wobble" causes the Earth's axis to carve out a circular path such that approximately every 26,000 years the axis will point to the same celestial location.

Proxy: a stand in or substitute. When direct measurements are unavailable, proxy measurements can be used to provide insight into past processes.

Qualitative: information describing the quality of something rather than the quantity of something.

Quantitative: information describing the quantity of something rather than the quality of something.

Quartile: each one of four equal groups that a set of data can be divided into.

References: a list of sources directly used as citations in a text.

Regression analysis: a statistical approach used to determine the relationship between two variables, commonly for predictive purposes. Linear regression analysis typically finds a "line of best fit" relating a dependent variable to an independent variable using the ordinary least squares approach.

Relevance: the significance, importance, or pertinence of the action or statement.

Residual: that which is left remaining when something has been removed.

Resolution: the interval between a series of measurements or observations in space or in time.

Respiration: the chemical process that provides cells with oxygen. Carbon dioxide is a byproduct of respiration.

Running average: a series of mean values calculated sequentially using a subset of the total possible datapoints (with a defined number of values = the averaging window) for each calculation. Each new calculation is done by shifting the subset of values used to generate the mean values incrementally one datapoint at a time through the series of data. The result is a series of mean values that smooth out variability in the original data set.

Scatter plot: a two-dimensional graph presenting discrete points, whose coordinates on the *x*- and *y*- axes represent two variables.

Scientific method: a process by which scientific information is developed and revised over time.

Spatial: relating to space.

Standard deviation: A statistic describing the amount of variation present in a data set.

Stock: the amount of something in a particular place at a particular time. Sometimes a stock is also called a reservoir.

Sub-decadal: spanning a timescale of less than one decade.

Temporal: relating to time

Time series: a data set that is collected over time. In a time-series graph, the *x*-axis is commonly time.

Topography: the elevation of physical features on the surface of the Earth.

Trend: the longest scale of change present in a data set.

Variance: a statistical term calculated as the square of the standard deviation.

Wavelength: the distance between two points in the same position of a cycle: peak to peak or trough to trough.

Index

Printed in the United States
by Baker & Taylor Publisher Services